MY LIFE AT AOL

By

Julia L. Wilkinson

1stBooks - rev. 3/23/01

Acknowledgements

Early employees of AOL used to joke about the "oral history" of the company: where it came from, who started it, and amusing anecdotes about employees and executives at all levels. I wrote this book because I wanted to share what it was like to work at this company which came of age in the dawn of the Internet revolution -- not from an outsider's, or even top-level perspective, but from the perspective of the many employees "in the trenches." I also wanted to offer a perspective of the Internet from one who watched it grow from obscure niche network to international sensation, and hopefully offer some helpful "back story" knowledge along the way.

As with many things in life, this book has taken me a lot longer and involved a lot more hard work than initially imagined. However, it has also been a good deal of fun, and I greatly enjoyed talking to the many people who gave me interviews, information, and suggestions.

I have many people to thank, starting with the AOL "old-timers" who gave me their insights about how it all began. Jim Kimsey, always candid and fascinating, shared his personal story as well as that of the company's. Frank Caufield gave his insights about "AOL's Forgotten Father," Bill von Meister, and the company's earliest days. With much humor and good nature, Marc Seriff shared invaluable stories about the immense technical challenges that faced the company from day one. My favorite AOL orator, Ted Leonsis, sat down with me to answer questions.

My sincere appreciation goes to Tom Brooke, for his consummate professionalism and timely advice.

I'd also like to thank my agent, Glenn Mollette, for believing in the book from the beginning.

Thanks to Chris Varner for his many leads, and giving me the first versions of the "oral history." I also extend my sincere appreciation to Duncan Champney, Janet Hunter, Jonathan Bulkeley, Lavona Rann, Robert Scott, Craig Dykstra, Bill Gorman, Bob Smith, Robert Seidman, Einud Stefferud, Vint Cerf, and Guy Kawasaki.

Others who gave me valuable feedback about their life in cyberspace: Calvert Deforest, Jackie Martling, Adam Curry, Shawna Koch, and Kevin J. Anderson.

I'd also like to thank the people who read and edited portions of the book, including Jerry Sweet, Signe Kavanagh, Magdalena LoGrande, Rebecca Thomas, and various members of the Sherwood and "ReWrites" Writers Groups.

Finally, thanks to my husband, Nick, for reading the manuscript, and for his support.

TABLE OF CONTENTS

Chapter 1: Is This a Career for You?... 1

Chapter 2: You May Ask Yourself, How Did We Get Here? 7

Chapter 3: Online Schizophrenia: The Many Faces of AOL................ 25

Chapter 4: Forty Weddings and a Funeral: The Virtual Life of
            People Connection 36

Chapter 5: Zen and the Art of Cyber-Emceeing: Celebrities Online 46

Chapter 6: Corporate Culture... 75

Chapter 7: Cybercity Ghetto 107

Chapter 8: An Awesome Tool; A New Way of Life 122

Chapter 9: Lifestyles of the Rich and Nerdy 132

Chapter 10: Leaving AOL: Unlatching the Golden Handcuffs 137

Appendix: A Cyber-Lingo Glossary: How to Speak AOL 139

Chapter One: Is This a Career for You?

My job was so free of interest, it would have made a great loan.

It was thus I found myself one Sunday in the summer of 1988, sitting on the sugar-maple carpeted floor of the one-bedroom Alexandria, Virginia high-rise apartment I shared with my sister, thumbing through the copious classified ad pages of Sunday's Washington Post. I was looking for that job description that would save me from my mind-numbing gig as a proofreader at a large D.C. law firm.

I'd seen the ad and it looked bizarre but intriguing:

"Writer/Producer

We're Quantum Computer Services, an innovative leader in providing interactive online services to the home computer market. We're searching for a creative, flexible individual to sharpen our competitive edge in the educational/entertainment market.

As Writer/Assistant Producer, you will write promotions and program online events, new services and newsletter material; coordinate a monthly events calendar; maintain live data base areas; and develop and produce online contests.

You may qualify if you're a real self-starter with natural promotional talents who loves details...thrives on deadlines...communicates clearly...and has a B.A. in Communications, Journalism, or Advertising. Some computer experience is a definite plus. If you have the spark we're looking for, let us hear from you now in writing...";

[illust. 1]

Writer/Producer? What was this, Hollywood? And online services? To the home computer market? What the heck was that? But it sounded intriguing. And futuristic.

I thought it was a very long shot, but figured what the heck. As the song says, "when you got nothin', you got nothin' to lose." I gathered up three of my best writing samples from my stint at the college student newspaper The Cavalier Daily, and sent them in with my cover letter.

And so it was that I got the job at a little company called Quantum Computer Services in 1988. What would happen to that company in the next several years would be amazing, even more amazing than what had happened to it in the several years before.

***

This stint as "Writer/Producer" was to be my first "real job." I was then working as a proofreader for a swanky law firm, and although it was fun to work

2

in an environment where people wore leather dresses and fur coats to work (remember, this was the 80s), I figured it was time to start using the creative part of my mind, if it was still there.

My non-illustrious collegiate career had yielded little in the way of good grades. But I did have a few interesting college newspaper clippings to my name, thanks to working for the Lifestyles department of The Cavalier Daily, and the fact that the University of Virginia was visited by a number of kooky and interesting characters.

One of my favorite interviews was with Rob Coles, who was Thomas Jefferson's fifth great-grandson through the female line. But the real clincher was that Coles was a dead-ringer for ol' T.J., the latter who was nothing short of a deity at "The University," the school he founded and which was so dear to him. (Thomas Jefferson requested that only three things be listed on his epitaph: "Author of the Declaration of American Independence, of the Statute of Virginia for Religious Freedom, and Father of the University of Virginia.").

Coles was the same weight, height, and build as Jefferson; was born in the same area, and had the same Albemarle County, Virginia accent. He had attended the University, but dropped out. When I mentioned to him one of my favorite Jeffersonian quotes about U.Va. – "a place where you were free to succeed or fail" – he said, with self-deprecating humor, "I guess I was one of those who chose to fail." But he was a success as an adult with a one-man show about Jefferson's life and thoughts.

So I felt that my interview with Coles, complete with a photo of his striking face, reflected positively on my abilities as a writer, which I felt were being thoroughly squelched by my stint as a glorified comma-mover at The Firm.

Another great assignment I had at the "U" was interviewing "Shoe" comic strip author and political cartoonist Jeff MacNelly. "Shoe" was short for "Shoemaker," the surname of his first editor at the Chapel Hill Weekly, and a cigar-smoker and sneaker-wearer just like the bird character in the strip.

"I didn't really want to call it that because I didn't want to embarrass him," said MacNelly, "but it just sort of stuck."

And did his ex-editor feel honored for having a comic strip named after him? I asked MacNelly for my piece. "No; I think he's pretty pissed," he said.

So it was with these nuggets that I hoped to get the attention of the Powers that Were at this mysterious computer company.

***

Quantum Computer Services's offices were located in a nondescript four-story office building in Tysons Corner, Virginia, a short ways west from that infamous behemoth of modern-day materialism, Tysons Corner Mall. I hadn't been to Tysons Corner for years, since when I lived in nearby Reston, Virginia as a kid. In those days, it was an ex-urban outpost, best known for its large mall, and a few car dealerships in between. Over the years, it had spawned more and more office buildings, chain restaurants like TGIFridays and Bennigan's, and stores like Marshall's discount clothing, Payless Shoes, and CompUSA computers.

I put on the best suit I owned, an inexpensive but passable brown-and-black checked tweedy number from Marshall's, donned my heels, and bravely headed out onto the Washington Beltway in my puke-green 1981 Chrysler "K" car station wagon.

My interview with Quantum was an exuberant experience. Unlike any other job interview, where you're asked the obligatory questions "where do you see yourself six months from now?" and "What are your biggest weaknesses?" this one yielded refreshing queries. (To that last question, I knew better than to say anything but "I can be just *too* organized; *too* hard-working" or "I am more detail-oriented than The Anal-Retentive Chef").

Here, I was asked, "What's your favorite tv commercial?" (the Veryfine juice ad where the guy's body looks like it's navigating roads like a human car) and "What's the last book you read?" (Fortunately for me I had wedged "Ogilvy on Advertising" in between Jay McInerney and Judith Krantz).

My first interview was with Christine Leberer, a diminutive thirtysomething executive with wavy strawberry blonde hair and a low-key demeanor. The pink-and-green inflatable palm trees on her computer told me she had a sense of fun. She had been hired from Viewtron, an early "videotex" (no "t") service started by Knight-Ridder, where the startup Quantum had recruited a few online service pioneers.

After meeting Chris, they asked me back for a second interview, where I met several of the software engineers, and some of the other producers, including one guy who explained how he would "hang out" at an online bar with his pet alligator, who he kept on a leash. I was just beginning to get that this place was headquarters to a whole imaginary world of role-playing, not unlike the "Dungeons and Dragons" game which engrossed so many kids in the late 1970s. (J.R.R. Tolkien's wildly popular book "The Hobbit," and subsequent trilogy "The Lord of the Rings," had spawned this world of fantastic Middle-Earthian characters and imagination).

A human resources manager sat me down and explained the salary: $22,000 to start. When I told him I was currently making only $15K a year, he asked, tongue-in-cheek, "how do you live on that?" And yet the salaries that Quantum was offering at the time were not exorbitant. In fact, to make up for that, the company offered something called stock options. When the HR manager handed me the stapled document explaining what would happen should this little privately-held startup ever "go public," I gave it hardly a glance, figured, "yeah, whatever"... but kept the paperwork safe in a file, just in case.

I wasn't the only AOL employee who knew from the beginning I was in for something different. "I was in a job I hated down in DC, and answered the old "put some magic in your career" help wanted ad in the Washington Post with the wizard on it," said one AOL developer. "I went to a job fair at the Sheraton in Tysons Corner, just up the street from AOL. I walked into the lobby, and was confronted with a six-foot-tall inflatable godzilla wearing a 'PC-Link' tee shirt. I thought, 'Cool! I could work here!' I then met with several people, all cool. I met the HR director, and she was wearing acid-washed blue jeans."

***

"Is This a Career for You?"

Twenty-two K or not, the truth was, this was the end of an era for me: a hash-slinging, mindless, low-paying string of crap jobs that are a rite of passage for most American kids whose parents aren't loaded and who don't want to be seen wearing the same pair of Toughskins the rest of their adult lives.

I had worked cleaning houses, cleaning toilets, and cleaning dishes; slinging grub at a bakery, deli, and Tastee-Freez; hawked wares at a jewelry store, five-and-dime, and two clothing stores; hostessing at two restaurants; delivering papers; filing magnetic tapes for a phone company, and finally, playing glorified comma-jockey as a proofreader at Debevoise.

College brought a brief respite from the world of paying jobs. I didn't want to jeopardize my brilliant academic career with time spent working. But this strategy didn't work much anyway, since said academic career was pretty much blown grade-wise after my first disastrous semester in Virginia's E-school.

However, in the fourth semester of my fourth year at Virginia, I came to terms with the harsh reality that a 4.0 GPA wasn't bloody likely. Also, I wanted a little dough for a trip I planned to take to Europe that summer, so I bit the bullet and took a job with a deli in Charlottesville. I just prayed that none of my sorority sisters ever patronized the place (a few of them did once, and I had no idea who they were. I guess you could say I was a less-than-model Greek).

The deli's staff was a hybrid of college students, locals, and transient adults living in Charlottesville. Once, while immersed in the subtle nuances of making the perfect Reuben, one of my co-workers (who was one of the transient adults) turned to another deli worker (who was a college student) and asked, deadpan, "is this a career for you, Tom?" Call it snot-nosed snobbishness, but that became our mantra-like joke while working there, and we'd often turn to each other in all mock seriousness and ask, "Is this a career for you?"

Exposé

For whatever reason, I have a long and sordid history of guys exposing themselves to me on the job. It all started with my very first job, a paper route for the Washington Star when I was ten years old. One of my customers lived at my "drop point," which was the apartment building where the Washington Star's truck dropped off the stack of newspapers every morning.

One afternoon, when I went to his apartment to collect for the paper, he answered the door (this was a rarity, since he was one of my "deadbeats" and usually pretended not to be home, even though I would see his peephole darken, a wary eyeball behind it). He invited me in while he got his checkbook, and I waited in the foyer while he wandered off. When he returned back to the foyer, he was not wearing any pants! Or underwear either. I got out of there as fast as I could after taking his money. The worst thing was I never knew when I would run into him again, since he sometimes came home when I was putting the papers on my cart from the drop point.

And then there was my disastrous brush with waitressing. When I was nineteen, I answered a window ad for "Waitress Wanted" at a pizza place a block from where I lived in Richmond, the summer of '84. The place was very slow —I only had a few tables all night. At the end of the night, as I was tallying the checks at one of the tables, the restaurant's sole employee that night, a Sicilian guy of about twenty-two, came over and stood next to me. He proceeded to unzip his fly and take down his pants. I can only imagine what he expected me to do.

I left in a flurry, and the place actually called me the next day to ask "why she no come back to work tonight?"

I mention this as a prelude to one of the most egregious examples of exhibitionism I would witness, courtesy of Q-Link, Quantum's online service. In the early days of Q-Link, the service had a very popular virtual nightspot known as "Bonnie's Bar," run by a real-life woman named Bonnie. As any good bartender would, Bonnie paid attention to her customers, and some of them, particularly the male ones, could get carried away. One of them actually sent her a series of photographs through the mail that would make anyone's hair curl. He'd mounted a camera at an angle above his bed, and taken several shots of himself masturbating. Let's just say this guy was by no means, as Seinfeld would say, "master of his domain."

But I'm getting ahead of myself. Those pictures of that guy jerking off were just one of the curious examples of human behavior I would witness via the strange and wonderful job I was to have at Quantum, which later became AOL.

Chapter Two: You May Ask Yourself, How Did We Get Here?

My first day on the job at Quantum, I didn't know quite what to expect. I'd had brief flashes of this "online service" during the interview – every employee had one or more (usually three) computers in their office, and most of them were signed up to the system.

The signon screen of Quantum's online service, which everyone called Q-Link for short, was an animated virtual marquee, flashing different colors all around.

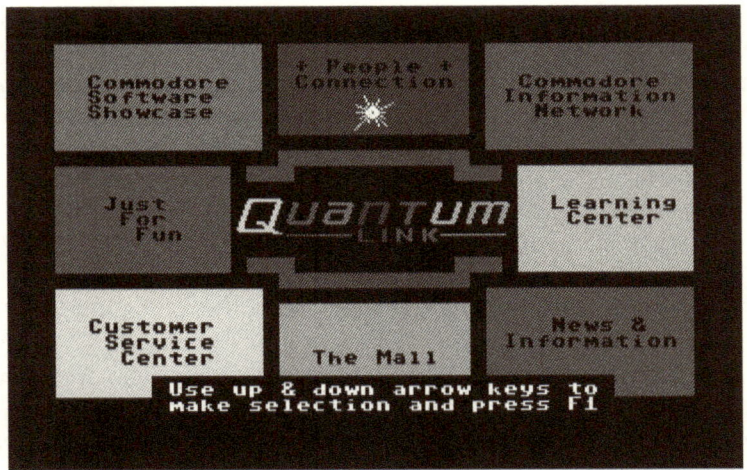

[illust. 2: QuantumLink Signon Screen and Main Menu, courtesy of
http://www2.ari.net/jpurkey/qscreens.html]

One co-worker described Q-Link as "cable t.v. for the computer." I was left alone with a Commodore 64, the computer which Q-Link was built to work with, to troll around the menus and explore. There were more nooks and crannies than you'd find in an English Muffin factory. There were areas with information and member posts about news, sports, recreation, health, you name it. There were games and clubs for people with similar interests. There was stuff about just about anything under the sun, only not in encyclopedic depth.

I was beginning to get a handle on just what this online service thing was. But it would be years before my friends and family would understand what I did for a living, much less use it themselves. Still, I had many questions. Just who used this thing? And how did it come about?

***

In the 1970s, the personal computer revolution was in its infancy. Most households didn't know what a personal computer was, much less own one. But the concept of interacting with a video device in the home was around. In this case, the video device was the plain old CRT (cathode ray tube) of the television screen, and the company that brought video games to homes was Atari Communications.

Ask a kid today what was the first video game, and they will reply, "Pong." Pong was the first simple "table-tennis" game, which Atari produced. It allowed a single user or two users to play a game of tennis, using two simple rectangles on the screen as "paddles."

But as the 70s came to a close, other uses for the CRT were in the works. What if you could retrieve information on that screen, and interact with it by typing? That would require a personal computer, and not just one...one that was linked up to others.

One day in 1986, I was hanging out in my boyfriend's dorm, when I watched him type a message on the bottom of his screen. "What are you typing?" I asked.

"I'm just talking to a friend across campus," he said.

"You mean he's on a computer, like you, and you're communicating?"

"Yep...isn't it cool?"

"But how are they connected?" asked Dave, a friend who was hanging out with us.

"Telephone wires," shrugged my boyfriend, as if he's been asked what color the sky was.

It was my first experience with modems.

★★

Jim Kimsey, now Chairman Emeritus of America Online, has said, "the story of AOL has been one of the most entertaining corporate stories in America because it's so bizarre."

Indeed, it's a miracle that America Online exists at all, when you consider that it's the ultimate result of three previous companies that not only failed, but also spectacularly exploded. In fact, had its founders been less tenacious, AOL would be nothing more than some papers in a cigar box, where its ancestors dwell.

So how did the company eventually become the number one Internet/online service in the world, not to mention, with its merger with Time/Warner, a giant media conglomerate?

It all started with a restless racecar driver. He and a young maverick engineer noodled around with technologies that read like a timeline of America's love affair with things electronic of the late 70's and early 80's.

The Mad Genius/Inventor/Promoter

The year was 1981. Video games like those made by Atari and Activision were in their heyday. A man named Bill von Meister had an idea: to create a product that would download video games through a modem connected to a phone line. Instead of endlessly going to the store to buy new cartridges, a cassette would fit into your game machine onto which you would download new games.

Venture capitalist Frank Caufield was interested. Caufield's firm, Kleiner, Perkins, Caufield and Byers, has been cited in WIRED magazine as one of the names that comes up "when you rummage through the IPO investment prospectuses of the hottest companies in technology," and was behind Tandem and the original Apple investment. "We thought this was a pretty clever idea," he said.

Caufield described the man behind this brainchild, von Meister, as a "mad inventor/genius/marketer/promoter."

He was the son of wealthy parents: according to Kara Swisher in "aol.com," his father was a godson of Kaiser Wilhelm II, and his mother was a countess. Per one report, the elder von Meister commercially developed the offset printing process still used by many magazines. Von Meister grew up in the horse country of New Jersey and toyed with the idea of entering the Foreign Service after attending Georgetown University.

"When I found out the starting salary was $7000 a year, I forgot about it," he said. He got married, moved to Europe, and raced cars for a while. (According to "old-timer" AOL employee lore, his passion for racing cars would crop up again later while courting investors for Control Video Corporation, when he sped 100 mph on a bridge span between Virginia and the District, claiming speed limits didn't apply there. "That was the deal; 'Give us more money or I'll do it again,'" joked early Control Video employee Mike Ficco about von Meister's proclivities toward speeding.).

"Business Could Be Almost as Much Fun as Racing Cars"

Then an opportunity came up with a friend of the family that would change the course of his life, turning him into a serial entrepreneur...and eventually change the course of corporate history. The friend was a Western Union Telegraph Corporation executive who complained of problems scrapping the company's old Air Force communications systems. It was in the process of switching from outdated vacuum tubes to transistors, and needed the old material hauled away.

Von Meister realized the tubes "had tremendous salvage value," according to Inc. magazine. "They had copper in the communications wires, the vacuum tubes could be sold for 10 cents a piece, and so forth." Since he was the only one willing to pay for them, von Meister got them, giving Western Union $750 for the lot. He'd found a salvage company that was willing to haul the stuff away for him in exchange for half of his profits on the parts. He made $250,000.

Von Meister "discovered business could be almost as much fun as racing cars."

"What do people do that's really challenging?" asked von Meister rhetorically. "They race cars, climb mountains, go scuba diving, and that's about it. Most really do live lives of quiet desperation. Well, entrepreneurship can be challenging, too. There are rules, but there ain't many."

The vacuum tube windfall was only the beginning. Since he now had access to top Western Union executives, he found out about a new product called mailgrams. Von Meister smelled another opportunity. At the time, the methods for large-volume users to send mailgrams was limited, and von Meister came up with a better way to do it.

He pitched Western Union an idea involving a central computer that would sort the messages by location and deliver them to a Western Union dispatching operation. Western Union turned him down, but gave him their blessing to try it on his own. So he went to a California company named Xonics which invested $1.2 million to develop the equipment. They quickly became the largest sender of mailgrams. Three years later, von Meister netted another $1.2 million when Western Union bought the idea back for $6 million.

Living well in a Great Falls, Virginia mansion, replete with lighted tennis court, heated swimming pool with automatic cover, and canvas-draped Ferrari Boxer in the garage, von Meister enjoyed the fruits of his ideas. He even had a racing engine put in his BMW M635.

He had about 2 acres and "a butler called Sargeant, who came around refreshing your drinks," recalled Ficco.

"Bill's most memorable characteristic was that he lived hard and well...loved good food and wine, always had the latest in audio/video stuff and other high tech toys," remembered Marc Seriff, former head of AOL Development, who met von Meister at GTE Telenet. "He definitely was a fascinating guy."

"Nobody will Pay You Enough"

"Sure this is all a game. What is it they say? He who dies with the most toys wins," Bill once said. His philosophy was that simply being an employee isn't enough: "Nobody will pay you enough. You have to do it out of capital gains."

"There's no secret; I just look at things and how they can be improved. Can you come up with an idea that will do something better, faster, cheaper? That's what I ask myself," he said.

Some of von Meister's other ideas included TDX, one of the first companies to supply devices that automatically send telephone calls over the cheapest routes. London-based Cable & Wireless PLC invested in the venture, eventually buying out Von Meister with $700,000 for his 24% stake, according to Swisher.

Another was one of AOL's precursors, a data-retrieval online service called The Source which Reader's Digest Association Inc. bought after 15 months. Built in the mid-1970s, The Source was the very first "online service"; so von Meister was in more ways than one the father of online services.

"The Source was really the company that first said 'there may be a business in people using online services in their home,' " said Seriff.

Precursor to Digital City

Von Meister met Marc Seriff around 1980, just after Telenet was acquired by GTE and became GTE Telenet.

"Being acquired by a telephone company is something you never, ever want to go through," Seriff has said. They worked together on a project called InfoVision, AKA Infotech, a concept which astonishingly mirrors the "Digital Cities" of today's AOL.

"It was to be a bunch of local online services around the country tied together with a national service. Sound familiar?" said Seriff.

"Infotech was an effort for GTE to go into local marketplaces, form partnerships with important businesses — TV stations, newspapers, cable TV systems, and put up local online services with a national umbrella behind it," he said.

But Infotech never went anywhere because the two couldn't convince anybody there was a business there. "Unfortunately, GTE never got it and the project died on the vine," said Seriff.

Digital Music: First True AOL Predecessor Company

But Seriff's acquaintance with von Meister would pay off. Bill left GTE, and Seriff became V.P. of Quality Assurance for Telenet, a management job that he didn't feel was a good fit with his personality. "Bill called me and offered the technical/operations department of his new company, Digital Music, the first true predecessor company of AOL."

"I jumped at it. Digital Music was the best technical idea I've ever worked on. The company's product was the 'Home Music Store.' The concept was to send extremely high-quality digital audio through a satellite and a cable t.v.

11

system into homes. Remember, this was before compact discs. The service would supply five channels of commercial-free radio and three to five channels of pay-to-record uncut albums; we gave you the right to legally purchase albums through this medium. It was really a great product!" said Seriff.

"You called us up, gave us your $5, charged it to your Visa, and in the middle of the night we hit the remote control on your tape recorder, and we would download the album onto your tape recorder.

"Great concept; technically it all worked like a charm; in fact we believe that somewhere, under some floor in the city of Huntsville, Alabama, our cable modem is still connected to the Huntsville cable system, because nobody ever found it," he joked.

The company had "excellent backing; we had our own transponder on a satellite disk; we'd received tentative approval from companies like Time Warner to supply digital masters," said Seriff.

It was launched grandly, which was to become one of von Meister's trademarks. "We had a big press conference in New York and they brought digital masters over in armored cars so we could show them to people," said Seriff.

Digital Music had some interesting investors: "We had the financial backing of the Osmond family – the good side of that was we had planned to build our studio and our uplink in Provo, Utah, which is some of the most beautiful country in the world. By this point, we had grown to four vice presidents and a secretary, so we were really on the move," he said.

Record Industry Rebels

Like all good things, Digital Music had to come to an end. And an idea this good was sure to threaten the tills of some very lucrative companies that thrived on consumers getting their music the old-fashioned way.

"One day we picked up Billboard magazine. In Billboard there was a full-page letter to the editor from the then-president of Waxy Maxie's, the local record chain, and he'd heard about this thing called the Home Music Store from this company called Digital Music. He wanted the entire retail industry to band up against it, because we were going to destroy the retail industry by bypassing it," said Seriff, adding facetiously, "which was true."

Within two weeks, all of their suppliers decided that maybe this wasn't a business they wanted to be involved in, so their source of supply was gone. Within a few months after that, the company was in a cigar box.

Leftover Bandwidth

But a serendipitous discovery while working on Digital Music planted the seed for von Meister's and Seriff's next endeavor: "Bill and I formed the nucleus of the team that started the next company, largely because one of the by-products of downloading music into the home was the leftover bandwidth at the end of the spectrum after we divvied up these video channels into these very high-speed data channels – 800 meg data channels for the audio," said Seriff.

They had a dozen 9600-baud channels. (Ninety-six hundred-baud were amazingly fast in those days).

They started thinking, what could they use those channels for? "We could give them news...the big thing we could do was download stuff. In fact, we could download video games," explained Seriff.

So von Meister and Seriff set off to create another startup, the one that would morph into America Online.

They went back to the same venture capitalists. "This is one of the things Bill von Meister was better at than any other human I've ever met, which was taking money from venture capitalists, burning it all up, and then getting more money from the same venture capitalists," said Seriff.

That's how Control Video Corporation, the predecessor company to AOL, was born. "CVC was created to try to capitalize on the then-really hot video game market. It was the first time video games really hit the stratosphere," said Seriff.

***

Although von Meister loved the challenge of creating companies and implementing ideas, as with many "creative people," management was not his forte. "He could be very frustrating to work for — he was master at raising money for his companies, a master at coming up with neat concepts, but not very good at managing the company to produce results," recalls Seriff.

(Some of von Meister's ideas were truly "out there." One employee from that era recalls him having a company that would match and sell human kidneys).

"He was a hardcore entrepreneur that, frankly, was probably more interested in the fun of doing a start up than the possibility of enormous success. However, I can honestly say, that in all the years I worked with Bill, I was *never* bored!" said Seriff.

Von Meister also paid close attention to trends. For example, explaining why he started Quest Communications Corp., he said the initial idea had been to cash in on the trivia craze started by Trivial Pursuit.

"If you go into a Toys 'R' Us, there's a wall of trivia type games," he said in a Washington Post interview in September 1985.

He also noted the success of '976' telephone services in New York, such as one run by High Society magazine featuring aural sex that von Meister noted was "pulling in something close to half a million calls a day," he told the Washington Post.

But his idea was not so titillating: he would fuse trivia games and dial-up services. The way the service worked was that you'd call an 800 number, listen to a series of multiple-choice trivia questions, and answer them by punching numbers on your telephone. You could win up to $100 for answering the questions right, and pay $2 a game for that privilege.

Although that company stopped operations due to undercapitalization, von Meister was undaunted and changed the company's direction, planning to make its mandate taking orders for companies that sold merchandise by phone.

So it was no surprise that von Meister started Control Video Corporation. It was, after all, was a company that meant to capitalize on another trend: the explosion of the video game industry.

Control Video would ultimately morph into AOL. But first, what ever happened to Bill von Meister? Sadly, he died broke, largely forgotten as the original founding father of AOL. He had a serious melanoma that claimed his life just six months after his diagnosis, according to Swisher.

His dislike for details dogging him, CVC's investors would sour on Von Meister, and he would eventually drift away from the company and its successor, leaving the reins to others more adept at managing.

(There had been many moments of tension along the way when the company's finances were tight – one employee's main memory of von Meister was him showing up one day in the office, poking his head in the doorway of an office, and joking "If a guy named 'Guido' shows up to repo the computers, let me know.")

Von Meister never got to profit from what the company he started became. His stock in the company became worthless, and he would not have stock in the company that followed. But although the world of magazines and newspapers has largely forgotten the role von Meister played, a few folks who knew von Meister in the early days will never forget him.

"For all his negatives, and all his positives, a lot of the success of AOL is a direct result of him," said Ficco. "His free-thinking ways... His high visibility stuff... I have to say, as a staff member, I liked it."

Ficco pointed out that von Meister's expensive taste and corporate flamboyance were lightning rods for talent: "Bill liked nice stuff. The offices were custom-made. Anyone at any substantial level had a nice office. Frankly, that's one of the reasons I was there, and one of the reasons why they had a lot of smart people in the early days. You could be proud to go to work; it wasn't a shack; it wasn't a warehouse with a lot of torn furniture."

"He was the ultimate salesman," remembered Ficco.

Ficco's first encounter with von Meister was memorable: He was coming to CVC to see about a job. "I parked my car in the parking lot, and walked past a black Ferrari Boxer. I observed the car...it was a nice car! The secretary brought me to the door; he was on the phone. He motioned me to come in. I was looking out the window, and I said 'nice view, I like that car down there.' And he held up the keys and said 'want to go for a ride?'"

Certainly, von Meister sold Ficco easily on working for CVC. He offered Ficco a job after their meeting, and when Ficco said he was not sure; that he was working for these defense contractors, etc., von Meister's sales pitch was, "I don't care what you're doing now; this will be more fun!" "And he was right!" said Ficco.

I never did meet von Meister, since by the time I joined in 1988, it had already become Quantum. When I interviewed AOL's chairman emeritus James Kimsey for this book, I lamented the fact that I was never able to ask the true "father of AOL" questions directly.

"Yes," he said. "Then you truly *would* have a book."

***

The Company Before the Company

So how exactly did Bill von Meister finally make a company "take"? Encouraged by what they saw around them in the hot-selling world of video games, von Meister's team and the investors were ready to take another crack at it. Caufield's firm bought into Control Video, as did video game company founder Bill Grubb, and venture capital firm Hambrecht and Quist. "It was a small investment," said Caufield; "about $400,000."

***

Atari's Heyday

"Atari had the Atari 2600, this dinky little game machine that had a 6502 inside of it, and was a god-awful machine, but people were buying cartridges as fast as they could put 'em out then," said Seriff.

"They were raking money in hand over fist. So we built a service called GameLine," he said. GameLine had a master module, which was a modem that plugged into the Atari VCS.

"We were able to put this guy on the dock at a time when 1200 baud modems were six or seven hundred dollars — unless of course you bought it from Apple, and then it was $1100," he joked.

A GameLine unit went for about $70. "You plugged it into your VCS, plugged the telephone in the side, stuck a battery in the little door, called us up on an 800 number to register, and lo and behold, a little keyboard would pop up on your screen that you could edit with your joystick. Through that you could use the GameLine magazine, or the catalogue that we sent you, which had listings of games in them – descriptions, listings, all of which had a little three-digit number next to them," said Seriff.

"You entered the three-digit number, we charged your Visa card $1.00, we download the game, as long as you left the power on, you could play this game as long as you wanted. We had contests, we would keep the scores, we were going to give away college scholarships and all kinds of good stuff," he said.

They went to the Consumer Electronics Show of 1983. "It was extremely well-received," said Seriff. "The CES of 1983 was a success beyond our wildest imagination. We took in orders for literally hundreds of thousands of these little devices."

"I learned a lot about retail in the six months after. But it was fun – this was a pure Bill von Meister CES.

"We had our own suite; we had showgirls that came up to our suite to talk to all the retailers. In order to attract the retailers into this suite we gave away gold bars," he said.

People-Poaching

They took in hundreds of thousands of orders. "It was great," said Seriff, adding, "Then all we had to do was to make it work. "

They began to put together a team. Seriff immediately thought of Telenet, where he'd had a good team.

The only minor problem was that he had a legal agreement with GTE that he wouldn't hire anyone from GTE. "However we discovered there was no prohibition on me *hiring* someone to hire people from GTE," he said.

"So that's exactly what we did. In March we got really busy developing the system. Not only did we have to develop the host system, the billing system, etc., but it turned out the video game industry at the time was really a garage industry," said Seriff. They had to make minor modifications to run these programs in this device.

They also wanted to do things like collect the scores for contests and things like that, so they went to the video game companies who had licensed them their titles for electronic distribution from the company and asked for a copy of their source code.

"When they stopped laughing, they would describe the guy that lived on a mountaintop – they'd never seen him – and he sent them ROM. As far as they were concerned, there was no source code, so we literally had to reverse-engineer every one of these cartridges in order to add our software. So that was a lot of fun," said Seriff.

They had to put in their own network, because these were proprietary modems. "We actually sold the rights to individuals to put in the local nodes as a business. It kept us alive for a while by collecting money from all these people," said Seriff.

The first person who bought one of those was a guy named Jim Kimsey. "He had some follow-on history with the company," said Seriff facetiously, "mostly to get even with us for selling him the local nodes."

Jim Kimsey: The Millionaire Soldier

Who is Jim Kimsey? Philanthropist, entrepreneur, Vietnam vet, straight from-the-hip-shooter –- any one of those descriptions of him would be correct. You will rarely meet anyone who filters what he says less than Jim does.

Jim Kimsey had always been a smooth, enigmatic figure around AOL. With all the press attention Steve Case gets, sometimes it's easy to forget that Jim was there when it all began. Once CEO of AOL, he eventually ceded that role to Case. I most remember Kimsey as the only guy to drive a Jaguar around AOL in the early days (and as I recall, they changed colors frequently).

Always a candid speaker, sometimes it was surprising what this top-level executive said in our company meetings. Around the time AOL did its deal with blue-chip computer giant IBM to produce an online service called "Promenade," Kimsey joked about being careful when you got in bed with the "big boys," lest they "roll over and crush you." He also counseled that we should tread lightly should a "peeing contest" develop.

He always seemed calm. Maybe that was because he had millions before AOL even started, and he would have millions should AOL ever go belly-up. Kimsey founded and served on the boards of a number of successful businesses in the Washington, D.C. area, from real estate development projects to restaurants. A Vietnam veteran and financial backer of an orphanage there, Kimsey was on the short list for ambassadorship to that country in early 1996.

He grew up in Washington, D.C., where he attended Gonzaga High School. "My Dad was a GS-3 [government worker rank], my mom was an Irish Catholic housewife," he said of his parents. After graduating from West Point and Georgetown University with honors, Kimsey served in the United States Army as an airborne ranger, where he ultimately became a Major. He won various awards for valor and service in the Dominican Republic and his two tours in Vietnam. If you called central casting and asked for the quintessential Vietnam War hero, you would get Jim.

Tour of Duty

Although it was over 25 years ago when Kimsey served in Vietnam, the country has stayed with him. One could say he is a surrogate father to a whole troupe of children: he founded an orphanage there and has continued to finance it all these years.

"The first tour I had was in South Vietnam," said Kimsey. "There were all these kids running around who were orphans....they were orphans we actually created," he adds in a soft voice. "We were in a very bad area. In fact my predecessor got wiped out." Kimsey was commander of an elite unit who was sent in to Duc Pho to replace him and his troops in 1965. Described as a "martyr" and "hero," the deceased leader had wanted to build an orphanage.

"So I ended up with this project," he explains.

During Kimsey's return to Hanoi in 1995, CBS's Bob Simon asked him why he built an orphanage in such a heavily Communist-controlled area. "Well, that's where the orphans were," he said.

Not only did Kimsey convince the US army to finance the place and build it where he wanted (which, by the way, would allow him to ask for more troops in order to defend it), he also convinced the Catholic Church to send four Vietnamese nuns down to run it. "It's unbelievable their ability to keep track [of me]," jokes Kimsey, who sends the institution money on a regular basis.

For Jim Kimsey, it's his legacy of a time of terrible carnage; his way of making amends to help the Vietnamese. As Bob Simon said of several veterans of the Vietnam War, "Good works have become their rosary."

"I'm sure I was responsible for hundreds if not thousands of deaths by calling in air strikes and artillery. I'm sure I was responsible for some of those orphans being created," he said. But, as he told Simon, "I think life is about moving on, being progressive, and doing the best you can."

Business Empire Started on a Shoestring

After the war, he returned to the D.C. area, where he bought a building on M St. NW with his $2000 in savings. He rented the top floor of an investment brokerage, which wanted to open a bar downstairs. They ran out of money, so Kimsey finished the job and opened The Exchange, complete with working ticker tape machine. The first Exchange restaurant was where the "Mad Hatter" is today.

Early Quantum and AOL Christmas parties were held at The Exchange, which at the time had moved to a location near the White House, and was an intimate little place. (I remember one such party, because I had a great time, and because it resulted in my becoming violently ill somewhere around the Washington Monument. Perhaps it was on the Washington Monument). A toy train was rigged to chug around the perimeter of the establishment's ceiling, and the atmosphere was warm and cozy.

This was quite a contrast to later vast and impersonal Christmas parties held at Union Station, in a room big enough to house an Inaugural Ball. Other parties would take up whole floors of the Westfields Conference Center in Chantilly, Virginia, where employees had to park in remote parking lots and be bussed into the hotel.

From there, Kimsey moved into other business ventures, including Bullfeathers restaurants, the Business Bank, and "a bunch of real estate projects."

The Accidental Babysitter

When Jim Kimsey left for a river rafting expedition down the Grand Canyon with his West Point buddy and former roommate Frank Caufield, he had no idea he would be taking the first step toward his main occupation for the next 20+ years. Kimsey had just sold a business, so he and Frank talked during the trip about what he might do now.

"It was all a big accident," said Kimsey.

Caufield told his friend, "We just made an investment in this little company all of about five minutes from where you live. Why don't you go down and check it out?"

Kimsey must have liked what he saw, because he became involved in Control Video first as a consultant, then as head of Manufacturing, head of Operations, and then as the number two guy.

Some put Jim's joining the company in more candid terms: "I don't remember what Business Week said re: how Jim came in, but it was bull," said Seriff. "Here we were burning through money as fast as we could print it, and Caufield called Jim and said go over there and figure out what they're doing. Jim actually entered the company as a manufacturing consultant. His job was to make sure the deals we were making with hardware manufacturers [were OK]...that was his *official* deal. His *actual* deal was, he was sent in to babysit and make sure we didn't burn through the money.

"He then became President and became the primary liaison with the moneymen, because we still had to go back to the well a number of times to the venture capitalist community," said Seriff.

The Company that West Point Built?

Kimsey's West Point connections, as a matter of fact, were a matter of some potential controversy.

Because of the initial tie between West Pointers Frank Caufield and Jim Kimsey, and the friends they had made in their class there, West Point became a tie that bound several of the CVC (and later AOL) board members. Jim Kimsey, Frank Caufield, General Al Haig, and Jim Andrus, head of a pharmaceutical company, were four out of the seven board members at one time.

Kimsey spoke of concern about the Wall Street Journal reporting about this, "because it looks like croneyism." He dismissed any allegations of wrongdoing or intrigue by pointing out there's nothing wrong with capitalizing on relationships if they'll help a legitimate business venture. "Well, but if my friends turn out to be good contacts... Frank Caufield is one of the best directors and venture capitalists in the world."

Many of the top-level executives in the company have been gotten by West Pointers, he pointed out – including [Chief Financial Officer] Len Leader, former head of AOL Networks Technologies Mike Connors; head of Marketing, Jan Brandt, and (now former) AOL International President Jack Davies.

Atari Implodes

In early 1983, everything was coming together for CVC. They had the product, they had their own host system, and they were supposed to start shipping in June of 1983.

But shortly after Jim started keeping his watchful eye on Control Video, a disastrous thing happened. "Around the first of May, 1983, Atari issued a press release and became the first company in the history of the U.S. to lose a billion dollars in a quarter. And they announced that basically the video game industry was dead," said Seriff.

"The company basically crashed," said Caufield of the market's implosion.

You Sold *How Many* Units?

Atari's troubles did not bode well for CVC's GameLine product, obviously. The tepid reception of the product with consumers was reflected in an early meeting with investors.

"I remember when they [CVC] launched their product, and there was a board meeting here in California," recalled Caufield. "There had been a nationwide launch, and much money had been spent in the course of it.

"Their head of marketing came up and presented the first sales report. They were trying to put the best possible spin on it. The guy reported proudly they'd sold 16,000 units. Everybody just sat there. I looked up and said, 'You'd have thought they'd shoplifted more than that!'" said Caufield. "Things kind of unraveled from there."

The disastrous events underscored the disparity between the successful demos of the product at the CES show, and what the market was willing to actually buy when the products were ready. "That's when we learned the difference between an *order* in the retail world, and a release to ship," said Seriff.

CVC finally did ship about 40,000 of the GameLine units. "We got about 37,000 of these units back," said Seriff. "We actually had, at peak, about 1500 customers, who were perfectly happy with the system. We still get, coming into the switchboard every once in a while a call from somebody who's bought one of these from a garage sale somewhere and wants to know how to turn on their service."

When it was all said and done, Control Video had a pile of 60-70,000 of these units. "The salvage value of each of these units was less than the cost to ship it to the salvager. The units ended up in a dumpster next to the company," said Seriff.

The numbers were sobering. And when certain boondoggling costs were factored in, were even more depressing. "That company burned through about $20 million in venture capital, and had a sum total of about $40,000 in revenue. It actually sounds pretty good until you realize that of the $40,000, $15,000 came from selling the hot air balloon they used at CES," he said. (CVC had purchased a hot air balloon to display outside one of the hotels at CES. "It actually looked like a joystick — it was amazing; it was gorgeous," said Seriff. It was also $15,000).

Brer Rabbit

"I ended up like Brer Rabbit with a tar baby," said Kimsey of the company at the time. "It kept getting in more and more trouble…it sucked up 20 million dollars. At one time I got a questionnaire from the SEC that asked 'have you ever been an officer or director of a company that went bankrupt?' I thought, I'll never be able to say no to that again. So I became very motivated to keep Control Video out of the tank."

CVC avoided bankruptcy, but went out of business.

MasterLine: The Last Stab

One might think that after this disaster, the principals involved would cut their losses and run with their tails between their legs to a "cog-in-the-wheel" job in corporate America. But they'd tasted blood, and weren't ready to pack it in.

"Before we closed up shop with GameLine, we made one more stab at it: same concept; a service called MasterLine. We caught on to the fact that video games were dead, but PCs were coming, and we adopted the same technology to a modem that we could plug into a PC," said Seriff.

"And that sort of began our partnering strategy. We went and we partnered with one of the newly spunoff RBOCs, [regional Bell Operating companies]

21

BellSouth – we were going to be able to keep all the service revenue. Really, really good deal. But, we almost blew it at the time when our contact at BellSouth, who was dealing with Steve [Steve Case, the young marketer who is now Chairman of AOL Time Warner, who we will meet shortly] called Bill von Meister and said he refused to deal with this whippersnapper anymore."

But the deal proceeded. "We were moving ahead, and then this funny little man named Judge Greene came in and said Nope, sorry, Bell Operating Cos. can't do this kind of thing; they're not allowed," said Seriff.

So once again, on the verge of moving into a business, von Meister and Seriff were thwarted. But by this time, they had a team. "Steve and Jim and I were sharing the management of the company…and we weren't quite ready to give up."

Enter Commodore; The Birth of Quantum Computer Services

Meantime, "all four of us formed a little consulting company on the side, so we could do things like eat when we were trying to get all this stuff coming together," said Seriff.

Kimsey began looking for strategic partners. He tried to take the PC/modem service idea to Apple, but the friction between John Sculley and Steven Jobs made that a bad bet. (Sculley was the former head of Pepsi Co., who Apple hired to be CEO. He and Apple's maverick, mercurial Steven Jobs had many clashes, until Jobs's eventual forcing out of the company).

Then, another player came into the picture. "Commodore computer was then probably the hottest consumer PC company in the world. And they had a corporate planning guy who was really interested in online services," said Seriff.

So Kimsey went to Commodore. "The Commodore guy and I got along. But he was too smart to buy it so I suggested that we start this joint venture to make the technology available," said Kimsey.

Commodore had found two companies they were considering for building an online service specifically for Commodore users: One was Kimsey and Seriff's, a small, not-yet-formed company called Quantum Computer Services, in Vienna, Virginia; the other was a company in upstate New York called PlayNet. PlayNet had devised technology for an online gaming service with a semi-graphic interface.

Ultimately, Quantum got the job, but with a hitch. Commodore walked in and said "you're the company, but in order to get the deal you have to go license PlayNet's technology," said Seriff.

So Quantum Computer Services was started, and Kimsey became CEO of that company. (Von Meister was no longer in that role…as Caufield explained, "it was probably mutual all around that the investors were tired of him.").

Turning Point

The Commodore deal was the turning point. AOL is now a multi-billion dollar company; PlayNet has been out of business for 13 years. "That was in the Spring/Summer of 1985. AOL officially came into being the summer of that year," said Seriff.

"We launched the service called Q-Link [short for Quantum Link]. It had little magazines and schedules, a Casino, and all kinds of good stuff," said Seriff.

Q-Link was "strictly for Commodore users. It was really nice for the operations people – and we were the operations people. We programmed during the day and ran the system at night. At 10 a.m. every morning, Q-Link closed its doors and didn't reopen them until 6 o'clock in the evening."

Because of Q-Link's limited production hours, the Quantum engineers didn't have to worry about operating an online service around the clock. "We had eight hours every day to do whatever we wanted with the system because there were no users on it," said Seriff. "It actually went pretty well by those standards."

But Q-Link had other problems. In the first of its many scaling problems (which would prove the bane of its existence after the company became America Online), the engineers learned the software they licensed would only grow to about 50 simultaneous users before it would tank. "We learned its reliability left something to be desired. I wrote a memo: "Congratulations – the system has been up for three consecutive days without a total system failure. That's a big deal," said Seriff.

The advantage they had was nobody cared about the Commodore users. "Nobody believed there was much of a market there, so we were able to proceed pretty quickly in building a service to a peak of 80,000 or 90,000 users, which for the time was pretty good," he said.

With all the struggles to keep the service up and running and grow it with its audience, it was easy for the Quantum team to sometimes lose sight of the importance of what they were creating.

"It was the first service to treat the PC as anything more than a model 33 teletype. It was the first time that anybody had looked at these PCs and said "Wait a minute! This computer is more powerful than the mainframe we were using three years ago. So we were the first to actually have software running in the PC cooperating with host software," said Seriff.

The Early AOL Vision

According to Kimsey, the whole vision of what has become AOL was never really the brainchild of Steve Case. It started with a guy named Clive Smith. Smith, who was an employee of Commodore at the time, was the one who suggested marrying the technology of the erstwhile online service "PlayNet" and

Control Video, and distributing them in Commodore boxes with Commodore machines.

"The core of all our software is PlayNet," said Kimsey, who had licensed it for CVC. But PlayNet was in financial in trouble at the time, and Kimsey needed to raise money for the company and get the partnership off the ground. "It was my job to get all the venture capitalists, get everything squared away with Commodore," he said. "The only person who did have money was Frank Caufield, and he was off on a trek in Nepal, walking the ridges." Eventually Kimsey got the money situation straight, and the fledgling Quantum-Link slowly took shape.

What exactly was Q-Link? That answer wasn't easy to give. One of my co-workers used to liken it to "cable t.v. for the computer." Another called it "an information and entertainment online service for owners of personal computers."

All I knew was that it made my computer come alive. It made computers — which I had previously only known via my frustrating days studying the Pascal programming language at college (where a missing semicolon could make or break you when running your program through a compiler) — fun.

Q-Link may have been an obscure product, but it was also a ground-breaking service. At a time when most "videotex" companies only offered plain-vanilla ASCII-text based information and chat, Q-Link's front screen sported flashy multi-color animation like a high-tech marquee. Its ground-breaking chat concept, which allowed people to create virtual "rooms," either public or private, and assign them names they thought up, made Q-Link's chat system the best interactive online experience out there.

And Q-Link contained all the key elements that AOL eventually would have: an interface that simplified the online experience, the large central chat area known as "People Connection," and special interest areas for information and bulletin board communication.

This was to become my home for the next several years.

Chapter Three: Online Schizophrenia: The Many Faces of AOL

My first job at QCS was writing online promotions for the Q-Link service. These were the equivalent of the little blurbs you see on the "Welcome" screen of AOL today. I wrote teasers for events like the auditorium conferences (favorite guests in those days were people like Commodore expert Jim Butterfield, and on the really rare occasion we would get 60s-guru-come-software enthusiast Dr. Timothy Leary. More on him later).

There were game shows that were variations on both popular tv shows and board games of the day, such as "Wheel Watchers," "MADvertising" and "reMODEM control," a spinoff of MTV's then-popular "Remote Control."

In a tradition that would become much more widespread in the decade to come, Q-Linkers often met "offline" for parties. People would go around introducing themselves in their online screen names, which were often "handles" like "Wolfhawk" or "MissJules," in the case of two of my fellow producers, which was how they were known by the others on Q-Link. My own first screen name was an odd choice in retrospect: "SlimeMold." I got it from a favorite cartoon by Matt Groening, who drew "Life in Hell" at the time, and was later best-known for creating "The Simpsons" on tv. He had drawn an organization chart of life forms, mixing things like "Republicans" and "right-wing fundamentalists" with "pond scum" and "slime molds."

I had the luxury of being able to create such a screen name because as a promotions writer I didn't need a public online personality. But that would soon change.

Quantum was branching out into other personal computer platforms. They'd recently signed a deal with Apple Corporation, and now they were planning to produce an online service for the Tandy personal computer, calling it, naturally enough, PC-Link.

The PC-Link People Connection department had an opening for producer, since its current producer was moving on to bigger and better things in the games development department.

"Worse than the Pentagon": Plucking the Apple

Although it was founded with the service for the Commodore 64 computer, the world of the Commodore computer had proved too limiting. The Commodore market was not doing well in the U.S., so QCS had explored other companies that might become partners. Jim Kimsey and Steve Case, a young man who was then a recent hire to the company, set their sights on Apple Computer Corporation, the then-successful purveyors of elegantly simple personal computers.

"We got that deal because Steve took an apartment in Cupertino and would not leave Apple until they gave him a deal. He showed up every single day," said Seriff.

Case History

Back when AOL was a distant number three in the online services market, before it was even named AOL, the President of Quantum Computer Services was a young, handsome, marketing wunderkind named Steve Case. A "regular guy" in seemingly every way except his extraordinary success at such a young age and religious-like belief in the product he was molding, Steve was an unassuming figure in the hallways. He was most often wearing office casual: a polo or button-down shirt and khakis. In fact, he was featured wearing this trademark uniform in a "Gap" khakis ad that ran in Fortune magazine in 1995.

Steve was born in Hawaii, like his parents. He and his older brother Dan were only 13 months apart, and they were quite a team. According to Business Week, when Steve was six, they started a juice stand using limes from their backyard. They charged 2 cents a cup, but lots of customers evidently gave them a nickel and told them to keep the change. "We learned early the value of high margins," joked Dan.

Later, they formed "Case Enterprises," which Dan describes as an "international mail-order company." They sold all kinds of things...like seeds and greeting cards via mail-order and door-to-door. "We made a fortune; tens of dollars," said Dan in the Business Week article.

Another pursuit of Steve's shows his early skill in the art of the spin, which he was to show later again and again at AOL. He wrote album reviews for the Punahou School newspaper. Dan says "He'd write reviews of albums in dinky student newspapers and write letters to record companies saying he wrote for the leading newspaper read by teenagers in Hawaii, which was true."

(This is reminiscent of how the company used to refer to itself in early press releases, when we were third in the marketplace in terms of subscriber numbers. It would inevitably be some sort of superlative that would render us number one in that particular niche, like "the nation's fastest growing provider of online services to consumers in the United States.")

Indeed, one of the biggest spin-saves I ever saw him pull was when he turned the souring of the deal with Apple Computer to work with Quantum on the "AppleLink: Personal Edition" network into an opportunity: AppleLink in essence morphed into America Online. Another big one was remarking, after AOL's network had been down an unprecedented 19 hours after an installation procedural error, that the outage showed us all how important, and indeed necessary, AOL had become to many of us.

After graduating from Williams College, Case worked at Procter & Gamble for two years. Per the "Business Week" cover story on him in April 1996, there he worked on the Lilt home permanent kit and a new product called Abound, a hair-conditioning towelette. "Towelette? You bet!" was the cheesy slogan. "It was a disaster," Case is quoted as saying.

His next job was working for Pizza Hut, searching out new toppings. In the evenings he played with his Kaypro computer and his first online service subscription, The Source. "I thought there was something magic in sitting in a hotel room and connecting to all of this," he said.

"Meet my brother. Hire him."

It was actually Dan (also known as "Upper" Case, just as Steve was nicknamed "Lower" Case) who had introduced Steve to the founders of Control Video, a company which provided delivery of video games for Atari computer owners over the phone lines. Dan's firm, Hambrecht & Quist, had a stake in the company, and Dan served on Control Video's board.

"If you read the Business Week with Steve on the cover, they told a story about how Steve came to work here because we met him at CES and all this kind of stuff," said Seriff. "OK, here comes the *real* story: Dan Case is Steve's brother. At the time, Dan was Digital Music's and Control Video's lead investor. He called us up one day and said, "I'd like you to meet my brother. He and his friend have just started a little marketing/consulting firm. Hire him."

Dan had just given Control Video somewhere in the vicinity of two and a half million dollars and was the lead investor in giving them five or six million dollars, so they weren't really in a position to turn him down. "Steve did have impeccable credentials," said Seriff facetiously; "he'd spent the last two years traveling from Pizza Hut to Pizza Hut sampling pizza, trying to come up with new flavors for the company. He was probably 25 or 26 at the time."

Still, the young Case was very enthusiastic. "I thought the business was interesting, but I was not blown away," Dan recalled in a Washington Post interview. "But I will never forget the look on Steve's face when he started talking about it all. He was so excited because he had finally found his place." That place was in a company which provided services not unlike those that had the younger Case spellbound in his remote hotel rooms, logging into The Source with his Kaypro computer.

Steve was taken under the wing of Kimsey, who recalls, "We hired [Steve] because Dan asked us to. He was a kid; he was a marketing intern. He'd been out of college two years...he was writing down how many Cokes they sold at Pizza Hut or something. I mean, it was not heavy-duty stuff. He wanted to do something a little more meaningful."

Although the press has sometimes put a spin of friction on the Kimsey-Case relationship (some of which resulted from Kimsey's quote in a Washington Post story about the company needing "adult supervision"), their relationship is more of an amicable father-son one.

There was, however, friction between Steve Case and a former member of the board, Douglas Peabody. "He got kicked off the board," said Jonathan Bulkeley, President of AOL in England. "The scoop was he liked Steve a lot but he didn't think Steve should be running the company. And he sort of voted against Steve for being CEO. Steve always held that against him and ended up booting him off the board."

Steve has a good sense of humor, and he would usually participate in the fun if AOL was sponsoring it. AOL used to host "beer bashes" every other Friday, so hardworking employees could blow off steam and chat. The atmosphere was very informal; according to one account, a customer service employee gave Case a "wedgie" at one such party.

Around Halloween time, many employees would come to work dressed up, and there was an annual contest where prizes were given out for the best costumes. (I am still proud of winning second place one year for my "Bridesmaid of Frankenstein" costume, circa 1992). Probably having been told numerous times that he looked like William Kennedy Smith (at the time of Smith's rape charge scandal), Steve came wearing green medical scrubs. (Smith is a doctor). He did bear an uncanny resemblance to the guy. (He looked a bit like Jay Leno, too, and his voice has always reminded me of Leno's, complete with a comedian's expert timing and inflection). But even if Steve Case originally came to AOL via a family connection, I don't think anyone would dispute that his own belief in the medium, quiet tenacity and sheer tactical brilliance has been a big factor in AOL becoming the huge conglomerate it is today.

***

Steve eventually wore Apple down, and a deal was struck where QCS would develop a network, "AppleLink: Personal Edition," a service for the Apple II, as a sort of compliment to Apple's private network. AppleLink: Personal Edition had its own magazine, with Apple's technical father, Steve Wozniak on the front cover of the first issue.

The good news was that the deal provided an instant infusion of cash. "The bad news was they were in the decisionmaking," said Kimsey of Apple's micro-managing after AppleLink got off the ground. Eventually Apple decided they wanted out of the venture, and breached their contract.

Ultimately "it cost us about five million bucks on that deal," said Kimsey. "I went to Sculley afterwards and said "Apple was the worst organization I've ever dealt with, including the Pentagon. It was the worst thing I could think of to say to him."

But the AppleLink service led to the creation of QCS's Macintosh service, and what would ultimately become America Online. "We were already working on an Apple II service and the deal with Apple kicked off the effort to build a Macintosh service," said Seriff. "That deal, like the first one, eventually went south. Much like the first and second one, the terms for parting with Apple were very cordial; they gave us a bunch of money and said 'don't bother us again.' "

Quantum capitalized on the contractual windfall by funding massive amounts of development. "It took us to the next step, and at the end of that process AppleLink Personal Edition became the first edition of AOL," said Seriff.

Tandy and the "Trash-80"

While all this was going on with Apple, QCS was scanning the computing horizon for other major partners and platforms. "Our good buddy Steve, the developer's friend, decided that he didn't have enough deals, so he went to Ft. Worth, corporate headquarters for Tandy," jokes Seriff in remembering the third online service QCS was to develop.

Quantum soon cut a deal with Tandy Corp. to create a similar service (then called PC-Link), for the Tandy and IBM-compatible market.

"Tandy was pushing the TRS-80 (affectionately known as the "trash 80" at the time), one of the worst PCs on the market, running an OS called DeskMate, one of the worst OS's on the market," said Seriff. "But they wanted an online service, and Steve convinced them that nobody could do it like Quantum Computer Services could do it.

"Steve came back between Christmas '87 and New Year's of 1988 with a signed deal in his hand. Tandy's going to bundle our service, they're going to put it with every computer they sell in Radio Shack, it's going to be on the retail shelf in Radio Shack. Just an incredible deal."

Quantum had a big party on January first to celebrate the signing of this deal. "That's when Steve did the 'Oh by the way, we need a golden master [a finalized version of the software] in June,'" said Seriff.

"We'd never worked on this OS before – the tools were virtually nonexistent. And so I turned to Huntsman, [Ken Huntsman, who was on the technical team under Seriff at the time], and I said, 'Huntsman, you have to have a golden master by June,' " he joked. Fortunately, they found a programmer who could do it. They shipped in time for the Christmas market, and that was the beginning of PC-Link, the next service QCS offered.

PC-Link also did reasonably well. But something else was going on in the hardware market. "We were sort of getting a reputation now for being a jinx," joked Seriff, "because it wasn't too long after that, Tandy started declining in selling their own computers."

They weren't the only ones. It was about this time that Commodore was in serious decline as well. Commodore, as it existed in the U.S., is no longer with us. "But it did us good," said Marc. "Part of the deal with Commodore was that they got warrants in stock in Quantum. When Commodore went down, their largest single remaining asset was their stock in AOL," he noted with a smile.

***

At this point in the late 80's, Quantum distinguished itself from other online services of the time, such as Compuserve, Delphi, and Genie (General Electric's online service), by their customization of each network to the computer that used it. For example, Q-Link was made specially for the Commodore and had the look 'n' feel of that computer; PC-Link's pull-down menus and screens blended in with Tandy's "DeskMate" interface. And the Apple II product, AppleLink personal edition, looked just like the Apple II, complete with similar fonts and graphics.

For the employees of Quantum, it was great to see the company branching out and expanding to so many different types of personal computer platforms. But there was a rub: running three separate online services – Q-Link, AppleLink, and PC-Link, meant you had to do everything three times. We tried to establish synergies between areas, as the three different People Connection producers would meet to develop a cross-platform game show, for example. But the duplication of effort was still there; each piece of text for each service had to have a separate "record number" generated, and they had to be separately plugged in to the service using the company-built tools on the "Stratus." The Stratus was a "super mini" computer that was like a mainframe, if not technically a mainframe.

Everything had to be done in triplicate. It was tedious and time-consuming, and eventually the company would see that a little consolidation was in order. Curiously, the event that set the stage for Quantum creating a consumer-friendly brand name online service of its own, rather than creating online services as showpieces or extensions of OEM's (original equipment manufacturer's) machines, was another deal with Apple that went sour.

Apple Redux

At about the time AOL buried the Commodore service, Apple came back into the picture. "They showed up at our door and said 'build us an online service,' " said Seriff. "We said been there, done that, no way. We literally spent a month trying to tell Apple no, we didn't want to do this again. They wouldn't listen to us. We learned later they were deeply in negotiation with both Compuserve and Prodigy and didn't want to work with them. It was actually the same people involved in the first Apple deal that wanted to come back for another round. They said all you'll have to do is give us an RFP [request for proposal] and you'll win.

"We said we don't have time to do an RFP. They said, OK, we're sending five people out, we'll talk to you, we'll write the RFP, you sign it," he said.

Finally, QCS capitulated and signed a second deal, a deal which, once more, would go south, but ultimately proved favorable to the company.

Even with these events, or perhaps because of them, Steve Case was still eager to expand the range of QCS's services. His thirst for deals seemingly could not be quenched. Indeed, had QCS been playing a game of Monopoly, their strategy would be to buy everything they landed on.

"We were trucking along with Q-Link and PC-Link and AppleLink Personal Edition, and Steve said 'We don't have enough deals,' so not too much longer, we launched a service called Promenade, which was a private label service for IBM." (Promenade would not be a success as an online service, among other reasons, because IBM's conservatism led them to strictly regulate the chat area, which had pre-assigned room names).

At this point, as aforementioned, Quantum was faced with the onerous task of running three separate online services, each with its own discrete chat area and files. "At this point we were mostly adding host resources," recalls Seriff.

Seeing the decline of the Commodore and PC-Link interfaces, in 1991, Quantum decided to "commingle" the separate networks AppleLink and PC-Link, and create one flagship service, which would be integrated under the name "America Online." Seriff described this undertaking as "massive," and dubbed it "the infamous commingling project, which was almost as bad as the infamous Internationalization project [that came later]."

Before the commingling, producers and managers at QCS were essentially doing everything three times; once for the Q-Link service, once for PC-Link, and yet another time for AppleLink. It was a very schizophrenic feeling, and could lead to errors of nomenclature.

The three services' chat areas, or "People Connections," had been self-contained vessels holding only members who used computers of the same ilk. Once the services were "commingled," users of IBM compatibles were thrown into the same virtual venues as that of Apple and Macintosh owners. The gossip and "us-and-them"ing that went on among the members was like something out of Peyton Place. But the diversity and expanded numbers added up to more dynamic discussions, and something akin to that key metric for such a service: critical mass.

AO vs. AOL

And it was a good thing that critical mass was being reached. With the renaming of AppleLink to America Online, and the commingling of the services, the whole company's name, logically enough, was changed to America Online, Inc.

At the time of the renaming, Quantum was still an underdog in the big picture of online services. So it was viewed by some employees as a move with a lot of chutzpah, if not arrogance, but I think a lot of us hoped that the service, and the company, would live up to its new name.

There was even discussion at the time over whether the new company's acronym should be "AO" or "AOL." My opinion was actually that "AO" would be more accurate as an abbreviation, since it was abbreviating two words – "America Online," and not three, which would have been "America On Line." (Although many's the time I'd seen the company's name butchered as "America On Line," "America On-Line," or worse, "American Online" in the press). But I had to admit later, it was a good thing they went with "AOL," because it had a ring to it the other abbreviation lacked.

I recall standing at that haven for employee information-sharing, the Friday Beer Blast, and discussing the merits of each name with Mac developer Duncan Champney.

"I prefer AO," I said, giving him my reason that America Online was two words. And AO didn't sound too bad, did it? Though I had to admit, it reminded me of my college sorority, Chi O.

But Duncan answered like there was only one way to go, "I think it should be *AOL*."

Even before the abbreviation debate, the company actually had an employee-wide contest to come up with the best new name for the service. They gave prizes for the top three names; one winning entry being "Odyssey."

But none of those contest entries were used. Steve's choice was America Online, and I have to say, I liked it. Though we may have thought it pretentious at the time, before too long, it would seem appropriate.

Beating the Big Boys

Even with all the deal-cutting and customization of interfaces, Quantum (and then AOL) was the underdog of the major commercial online services for a long time. Compuserve, which had held the title of largest commercial online service when I first started at AOL, was rapidly eclipsed by a joint venture between IBM and Sears called Prodigy.

These backers of Prodigy had deep pockets, and they were digging deeply into them to market their baby. In the early '90s, Prodigy ads proliferated in direct mail, in magazines, and on television. And yet, even with all its funding, critical mistakes like overly conservative censorship policies and a diminished focus on e-mail, bulletin boards and chat kept Prodigy from living up to its name.

Could it be that not having such vast financial resources actually *helped* QCS? Jim Kimsey thought so. "These companies spent a billion and a half dollars, and they had less than a million subscribers," he would recall when I spoke with him in 1997. "I burned $10 million in venture capital; we have over eight million subscribers. The reason is we didn't have the resources, therefore we had to use our wits; we had to leverage everybody," he said.

From the earliest days of AOL, when the staff "huddled in an old basement," to today, with over 26 million subscribers, the "story of AOL has been one of the most entertaining corporate stories in America because it's so bizarre," Kimsey said.

Paul Allen and Microsoft

It wasn't just big Goliath companies like IBM and Sears, and later MCI and News Corporation, who funded the struggling Delphi online service, that were taken on by AOL's comparatively tiny David. AOL fended off the likes of none other than Microsoft and its number-two founder, Paul Allen, one of the richest men in the world and a big investor in high-tech startups.

"I'm not gonna like this little nerd; he's gonna piss me off," Kimsey thought in his characteristic soldier's style when he was traveling to see Bill Gates, CEO of Microsoft Corp. But it turned out he was wrong; Kimsey found him to be "very personal, very straightforward, very nonpretentious," and "smart as hell."

It was Gates's sidekick, Allen, who rubbed Kimsey the wrong way. "Paul was Gates's boyhood friend. I'm not a big fan of Paul Allen. Paul Allen had all these hangers-on; a big contingent. I didn't care for him at all; and I double-A sure didn't want him on my board," said Kimsey in my interview with him in his D.C. offices.

"In fact I said something to the [AOL] board to the effect of "If Paul Allen didn't have six billion dollars, I wouldn't walk across the street to say hello to him."

[Then-fellow AOL board member] Al Haig looked at me and said, "You know what? Because he has six billion dollars, I'll walk across the street to say hello to him," recalled Kimsey.

Steve Case got Allen all revved up about the stock – and Allen bought a lot of it. It slowly became clear that Allen was gunning for a controlling portion of AOL's stock, so AOL sought to adopt a "poison pill" strategy, which is meant to fend off such an attempt. A poison pill is basically a scheme in which any takeover attempt immediately allows the company to issue millions of existing shares to existing shareholders, making it much more expensive to acquire the company.

"The day we did the poison pill, he'd already bought through it greater than 20%. Well, we didn't know it. There's never been a case, save one, when a poison pill was triggered," said Kimsey. But at the time it happened, nobody knew he'd already triggered the pill.

"Those white boys must be talkin' 'bout money."

"I found out when Frank [Caufield] and I were in the New Orleans Jazzfest," said Kimsey. They had to set up a new board meeting to reset the pill at 25%, since Allen had already acquired 20%.

"Frank and I went off to find some pay phones. The racetrack where they have the Jazzfest in New Orleans is in a bad area of town. There was an old store on the corner where they had pay phones. It turns out it was a black neighborhood, and these were the community phones; that's what everybody uses.

"Frank and I were on there, and I'm conducting it, and Frank's on there listening. Of course, a line starts forming. They said 'those white boys must be talkin' about money.' Finally one lady says, 'I've got to call my mother. I have *got* to call my mother. How long you boys gonna be on that phone?'

"I wish we could get a picture of Frank and me on the phone, with a line of black people on either side of us, trying to reset the pill so we could fend off Paul. That's a true story, and there are a hundred of these," said Kimsey.

The board did succeed in resetting the pill and fending off Allen, but he was to take a form of revenge later. Allen eventually dumped all his shares unceremoniously at once, a move which caused the stock to drop some seven points in one day.

But the real loser in the affair was Allen. "I understand he made an economic mistake," said Caufield. "He made a lot of money on his shares, but on the other hand, if he'd held them, he would've made five times as much. I think it was partially he wasn't interested in the situation if he couldn't control it, and two, he might've lost some faith in the prospects of the company," he said.

A Long, Strange Trip

Most of the people you talk to from AOL will tell you they're surprised at how large the company has grown.

"It's been a wild up and down ride," acknowledges Seriff. "If you'd said there would be this many people in the company we'd have laughed at you. Certainly if you'd described that 100,000 simultaneous users was something that would be behind us, we'd have laughed at you. And if you'd told us a quarter of a *million simultaneous users* was anything we'd ever need to worry about..." he mused, adding, "I still remember being in Ken Huntsman's office and couldn't conceive of even having 32,000 simultaneous users."

Ho Chi Minh Strategy

Frank Caufield felt differently; he'd anticipated the success, but not the rate of it.

"I adopted a Ho Chi Minh strategy," said Caufield; "to keep a low profile and fight for decades...keep the overhead down. In the early days, it was always a hobbyist/gearhead kind of market, and small. If we could establish enough of a revenue base and corporate partnerships, we could last until this took off," he said.

Sooner or later this would take off, he figured. "I had no idea if it would be one year or ten years. When it finally did happen, the timing wasn't surprising, but the momentum and rapidity was."

***

But if the cast of characters was large on the business deal side of AOL, it was nothing compared to the wide scope of gurus, pundits and celebrities that we, the workers in the trenches, were facing.

Chapter Four: Forty Weddings and a Funeral: the Virtual Life of People Connection

My first taste of the Wild Online West that is People Connection came the first night I signed on to the Q-Link service. Chris, my first boss, walked out to the employee parking lot with me and we hauled the parts to a Commodore 64 computer to my K-wagon.

That night, I got the thing plugged in and working on the living room carpet of my Hunting Towers high-rise apartment, just a stone's throw from the Woodrow Wilson bridge into Maryland. I crouched on the floor, cross-legged, listened to the raspy sound of the 1200 baud modem connecting, and watched as the hopeful word "connecting..." displayed across the screen in large font. (The sound of an old-style dialup modem connecting is disconcerting; there's a two-tone pitch preceding the long crackly hissing sound of the modem connecting to the host computer; many first-time modem users reported to customer support that their modems were "broken" after hearing it).

I tabbed over to the "People Connection" area and there this otherworld unfolded. It was like a typed-in party line; a CB radio with words. People who were all across the country were typing in their thoughts, jokes, even their own online "shorthand" of hugs, kisses, and winks. I had so much fun, I stayed up until 1 a.m. and crawled into work bleary-eyed the next day. My boss Chris joked with me about looking like I'd been in People Connection most of the night. But I think she was secretly glad I enjoyed it so much. I was going to be programming events for these people with the new job I'd be taking on after just three months at Quantum, PC-Link People Connection producer.

Randy Dean, the young UNC Chapel-Hill graduate who was my predecessor as PC-Link producer had already established a name for himself as "Tandeano" online – a combination of "Tandy" and "Deano," his nickname in "real life." He was moving on to a job in the Games department; we used to kid him that he was the real-life counterpart to Tom Hanks in the movie "Big," who basically got to play with toys all day in his job as "vice-president" at a major toy company.

Randy sat down with me to brain-dump his responsibilities. He had a mellow, good-natured disposition, and he took the job in stride.

"OK, first we have to create a screen name that you'll use as your online persona," he said.

"I guess SlimeMold isn't professional enough, huh?" I said.

"Nope."

"Well, how about 'Clueless?'" I asked, tongue in cheek. It was how I felt at the time.

He stared at me for a second too long. "Nope."

He was not amused. I better come up with something fast. Then something came to me. "How about 'Tangent'?" It was a combination of "Tandy" and "gent," and it also had a separate meaning as a word on its own. Plus, it was androgynous, and because women were still vastly outnumbered by men online in that time, I wanted to have some cloak of mystery about my persona.

"As in 'off on a...'"? he asked.

"Yeah."

"Tangent it is," he said, and typed in the word on the "create a screen name" form, saving it for posterity. (In one of those curious twists of fate, the screen name that had once been famous in that small niche world was eventually used in one of AOL's marketing pamphlets. I had given the company permission to use the name in this way, after I no longer used the name much myself. So when screen name "Tangent" was finally attached to a face, it was that of a fortysomething bald man with a black beard and glasses).

And so my first four years at AOL were spent working in the interactive area of the service known as "People Connection," a.k.a. the "chat" rooms. I started as producer for the PC-Link service, which at the time was Quantum's only service for the IBM-compatible "PC." (Why they call IBM-compatible computers "PCs" and Macintoshes just "Macs" I never understood; aren't they all theoretically "personal computers"?).

I had been in Q-Link's People Connection, but only as a participant. Now I would be creating live online entertainment events for these people...customers I would never see, but be able to communicate with online at night. And I would have a public image of sorts. Wouldn't I?

Could Andy Warhol Have Imagined This When He Said We Would All Get 15 Minutes of Fame?

Perhaps the best-known online "personality" in those days was a woman who went by the handle "Cupcake" on Compuserve. (Compuserve didn't have "screen names"; their usernames consisted of techie-like numbers and letters. However, you could create a name in their live chat area, known as "CB" — for Citizens Band, like the radio voice-communication truckers used).

Q-Link's first online star was "BonnieB3," of Bonnie's Bar, one of the key watering holes on Q-Link. The real Bonnie was a petite, attractive woman with wavy brown hair, who often dressed in jeans and cowboy boots. When I first met her, she was visiting the Quantum offices (she worked as a "remote host," from her home computer) and talking excitedly about the antics in her bar the night before. She is the one who showed me the photographs an overzealous fan had sent her through the mail of himself, er, not quite mastering his domain.

How did an online bar work? No, a hand doesn't actually come out of the computer screen and hand you a drink (though that would be a "killer app"); but you are pretending to be drinking with people and you role-play drinking and hugging others, with online gestures like the mug, which might look like this: {(_)

Hugs can look like this: { } or ( )

...or the double hug: {{ }} or (( ))

And even triple hug and beyond, for extra-familiar faces or special friends: {{{{{{ }}}}}}

Sometimes hugging would get out of control, and people would throw out absurdly long lines of parentheses made to resemble extreme hugging, like this:

{{{{{{{{{{{{{{{{{{{ Tangent!!! }}}}}}}}}}}}}}}}}}}

As producers, we sometimes worried that new members would be put off by seeing all this emphatic hugging going on among the old-timers. And there were more and more new members every day.

The Online Host-as-Celebrity

Being a high-profile People Connection host, like the legendary "Lavona" who ran the long-running "LaPub" on AppleLink (and later on America Online when the three services merged), brought with it a unique set of both blessings and curses.

Recalling her days as the "Pubtender" of this establishment, Lavona said, "One of the strangest [things] happened the week after the man who was to become my husband made an online public announcement that he wanted to spend the rest of his life with me. In a folder created by people to congratulate us, someone, as yet unidentified, posted fictitious 'facts and comments' that were easily disproved, about supposed seedy activities. Unfortunately we were away from phone lines for over a day and the world had seen the posts before we could comment on them or see that they were deleted.

"The screen name used to post the comments was new, yet did not exist when checked, which means that someone created it, posted and deleted the name. We have to this date, over five years later, not a clue as to who, or why that little bit of viciousness showed up," she said.

"On the brighter side, when financial problems outside of our control occurred and the word got around, people that were also among the volunteer workers started sending us, as individuals, unsolicited checks 'to help through the rough times.' Such an outpouring of love and unselfish giving is rare on this earth and even more so when you realize that some of the people had never met with us offline. When anyone asks me if friendships made online are real, I've many reasons to say emphatically yes."

Marriages Across America

This was where I witnessed that strange and wonderful other-world where things happened but didn't happen; where the shyest people became virtual stars, and fast friendships were made overnight. Marriages may have been destroyed, but my experience was that many more marriages were made than ruined. Just among my staff alone, I witnessed several personal relationships and marriages occur, so you can imagine how many there were among the general membership.

One People Connection room host met a new boyfriend in LA, and moved there from Ohio to be with him. Several months later I heard they had moved to Indonesia, and I received a wild e-mail about her exotic life there, with gekkos crawling all over the walls of her airy home.

Another regular, a trivia host who was known for his amazing ability at winning in trivia rooms himself, seemed the epitome of the family man. He and his wife were both involved in hosting People Connection rooms, and one of their daughters even took a crack at hosting a teen chat.

A few years after I stopped working with him, I heard that he'd left his wife to live with a woman he met online who lived in Alaska.

Still another host, a single Radio Shack employee with a very sweet disposition, wound up meeting the love of his life through the chat rooms. She was about fifteen years older than he, but he sent me a long tale of their cross-country car trip together, complete with the storybook account of their first face-to-face encounter.

***

In the early days of People Connection we flew a lot by the seat of our pants. When I first inherited the PC-Link People Connection realm from Randy Dean, the area's previous producer, he and my then-boss Chris had cooked up a whole metaphor for the area: it was to become a virtual hotel.

He and Chris felt this was a natural metaphor, because People Connection was already based on a "rooms" model. Then as today, when you entered People Connection, you were placed into a "lobby," which is kind of a way-station in cyberspace; a default room where people entering the chat service would plunk into. Then, to change rooms, you could choose from the menu of rooms available, which were rooms that had been created by other members of the service. Several of these were "hosted" rooms, which volunteer members ran with a theme in mind, such as Trivia or Bonnie's Bar.

Or, you could create a "private" room, which would not be visible to other members by the room list, but which you could invite other members into by instant messaging or emailing them the secret room name. (Occasionally a member would accidentally drop into such a room because they had inadvertently created a room with the same name. Depending on what was going on in there at the time, this could be embarrassing for the members.).

So the rooms, that were currently just "rooms" in cyberspace, were to turn into virtual hotel rooms in a cyber-establishment. Some of our regular rooms, such as the Lobby, fit this model well; others wouldn't so well. But a trivia room, teen chat room, or matchmaking room were less likely to survive this metaphor.

I was skeptical about the whole scheme from the start, but I didn't voice my opinions right away. My boss and I created an online hostess character known as "Gigi LaVrum" who was to welcome people to the Inn (which we dubbed the "Vienna Inn" after a real-life chili-dog-'n' beer dive in Virginia, renowned for the rudeness of its serving staff, where Quantum employees liked to congregate. Why we thought the Vienna Inn would have any meaning for anyone outside of Vienna, Virginia, I have no idea).

Well, the nice thing about working in an interactive service is you know right away when people don't like something. There were few posts on the hotel's designated boards, but those that did bother to say anything showed us how darn baffled they were. It was usually something succinct, like "Huh???" (People in cyberspace are prone to punctuation hyperbole). Or, "What's with this hotel thing?"

One day after a couple weeks of this charade, I walked into my boss's office. "About this hotel..." I started.

"We need to kill it, don't we?" She read my mind.

"Yeah," I laughed. And that was that. The Vienna Inn became good ol' general, free-for all People Connection again, and life continued apace. If there was a lesson to be learned there, I think it was that sometimes it's best not to mess with a good thing. (Or in other words, if it ain't broke, don't fix it).

Another thing I got out of it was that people's imaginations were already turning these "rooms" into all kinds of virtual destinations, from bars and beaches to wedding chapels, divorce courts and funeral parlors. We didn't have to supply a lot of fancy metaphors, nor should we necessarily constrain the possibilities they were already discovering on their own.

"May I have a Rheum, Please?"

In those days of the late eighties and early nineties, you could list all the rooms in People Connection on any given night in a couple clicks of the mouse; one or two screenfuls of information. The members could always create their own public or private "rooms," which held up to 23 people each. Part of my job was to decide what sort of "hosted" or "sanctioned" rooms we would offer.

There were usually one or two "pubs," which were virtual watering holes where a bartender/host would serve "beers" — which looked like [(_) or {(_) 's.

It was in one of these "pubs" that I was (virtually) drenched in champagne in a welcoming gesture by the remote staff of what used to be AppleLink People Connection. This was one of the rituals with which they "broke in" new AOL employees. The three People Connections of Q-Link, AppleLink and PC-Link were being consolidated under one management, and I was to take over for the previous AppleLink producer who was moving on to another job.

Another rite of passage of sorts was the jello dive, in which you took your turn in line to get a running start, then swan-dove into a huge vat of jello. (Of course you weren't actually jumping into a vat of jello — you were pretending to; but many "Linkers" described their online activities as though they actually happened in the physical world. This could be confusing for many who dwelt only in the offline world). This was done as a series of choreographed actions, usually delineated online by sets of colons, like this:

::Tangent gathers strength:::
:::starts running toward jello vat, gathering speed::
::jumping high into air::
::SPLAT!!! Landing face-first in jello:::
::mm.....licking jello off face:::

Well...you get the idea. Hopefully.

Programming for a Brave New Medium

If the initiation rituals were bizarre, the actual work involved in planning events for the people out there in cyberspace could be more bizarre still. You were treading on new territory, where there were no rules and you had a great deal of freedom to try new things.

And try new things we did. And learned many lessons.

The Online Kidnapping

One wacky night on Q-Link, a caper unfolded that caused excitement and took days to play out. Some roguish patrons of Bonnie's Bar decided to kidnap its lovely bartender, "BonnieB3." Bonnie was simply missing from the service for a while, and it caused a lot of rumblings and conjecture, and generated many posts on message boards.

The two People Connection producers, one of whom was working for the company when this "kidnapping" played out, thought it would be a fun idea to try to duplicate this kidnapping on the other services' People Connections (which were AppleLink's and PC-Link's, at the time).

It would play as an online mystery, we decided. Or we hoped. The victim would be named "Julsie," which was close to the screen name of one of the People Connection producers. Although it had the added benefit of being close to my real name, too.

What did we learn? That you can't force organic online events. People didn't get excited by the concept, since Julsie wasn't a real personality who had been hanging out online, and had connections and friendships, people didn't know what to make of it. These things have a life of their own, and the members will be the ones to generate them.

The Comedy Club

Being a fan of comedy and comedians, I thought it would be fun to have an "Open Mike Night" where aspiring comics could test out their jokes online.

We sought out local talent in the D.C. area, and I even had a good excuse to go to a Jerry Seinfeld show at the Warner Theatre in D.C. After he did his act and "Sinbad" came onstage, Jerry strolled out into the audience and took a seat just a few rows in front of me. I wanted to hand him my business card with "I want to be your milk monitor" on it. (Jerry had a bit back then about how he wanted a milk monitor, with a white towel draped over his arm, to stand by his fridge and do nothing but warn him when he was running out of milk). But I lost my nerve. So much for Jerry Seinfeld on PC-Link.

While the concept of a live comedy club in chat didn't work well, because the subtleties of tone, vocal inflection and timing didn't translate, we did have a popular text/bulletin board area on the service called the Comedy Club, where we pioneered things like "Punchlines" contests. Contests like these are now successfully done in such virtual places as the "Hecklers" area on AOL.

The Murder Mystery

"You should do an online murder mystery." If I had a nickel for every time I heard that from other employees at Quantum, I'd have…about a buck fifty. But I was skeptical. It sounded good in theory, but how would it actually be implemented? And more important, received? But unfortunately, once again I did not listen to that nagging inner voice, and we tried to do a murder mystery online. We posted the basics of a plot, which involved a "crooked cop," and hoped that members would guess at the whodunnit, or at least have fun guessing.

But the culmination of the event, held in a regular chat room, was a cacophony of typed-out guesses, jokes, and catcalls. Many people enjoyed themselves, but Sherlock Holmes had nothing to fear from us.

Still, if not all of our events panned out, we were not afraid to try new things, and many things we tried were successful. And AOL producers were usually very good at letting the their various communities, and hence the AOL community as a whole, grow organically, something that I think helped AOL's overall community to flourish.

The Treasure Hunt

Treasure hunts were another online event that were suggested again and again. And they were done again and again, by me several times in fact. And people enjoyed doing them, too, for a while.

How did an online treasure hunt work? You hid the "eggs," or "clues," as hints on message boards scattered throughout the whole Q-Link or PC-Link or AppleLink system. The hint could be in the subject of the posted message, or, if you really wanted to make things difficult, in just the text of the message.

If that sounds like it would be an impossible task, to track down even one much less 10 or 20 online hints on the vast AOL message boards of today, you're right. But remember the service was smaller and more intimate in those days. Treasure hunts could also be confined to smaller areas of the service, and might still work today in a small enough forum.

Not surprisingly, Easter was a popular time of year for "egg" treasure hunts online. And there was another aptness to calling it that: the term "Easter egg" was used by software programmers to refer to a hidden feature or graphic placed in a program. This was usually something the programmers themselves put in just for fun, or maybe to alleviate their boredom.

(One infamous "Easter egg" on AOL consisted of a naked cartoon butt popping out of the door of a psychedelic 60's bus when you clicked on it. This was on the front screen of AOL's erstwhile "Road Trips" area, where a member could conduct his own "tours" of the web, broadcasting a number of web pages in an upper window, while he and his friends chat about them in a lower window).

The Interactive Novel – Storyline

The interactive novel was another pet concept of users and Quantum employees alike. The idea was like that old game, "whispering down the lane," where someone starts a story, another person comes in and continues it, a third person then adds a plot twist, and so on, and so on.

We did this one a couple times as well. I think nobody would disagree with me that they can be great fun, but too many cooks can make for one strange literary brew.

The Online Wedding

What is an online wedding? This excerpt from the erstwhile Q-Link magazine, the "Update," attempts to explain this unique phenomenon, in a depiction of an online chat room conversation about such a virtual nuptial:

RJScott: Hey, everyone! There's a Qwedding in Bonnie's Bar!

EileenC: ::heading to Bonnie's::

YOU: A Qwedding?

RJScott: A couple gets married online…

YOU: You mean for real?

RJScott: For real…well…they think so!

YOU: How could you marry someone you met online?

RJScott: ;) (a wink)

The online wedding (which sometimes led to an offline wedding) was often the result of the expedited relationship unique to this new medium.

The Vanity Screen Name Auction

According to an early AOL "Update," "when Quantum's staff first developed Q-Link's software, little did they know the impact of anonymous 'screen names.'" Instead, the developers had been concerned about protecting members' identities on "what they hoped would be a large, national telecommunications network."

Simple screen names would set Quantum, later AOL, apart from the other commercial online services, like Compuserve and Prodigy, where arcane, technical and hard-to-remember IDs, like "drkf1897," would dampen the member experience, something AOL was so proud of. The staff rightly assumed the ability to create their own "screen name" would be more personable and entertaining.

Over the years, good screen names have become a commodity, as all the simple, and most desirable, ones (as in all first names, like "John" or "Susan," and even most first-name last-name combinations, which were short enough to fit into the then ten-character limit) were already taken.

Good screen names certainly carry a caché. I've never seen an AOL employee as happy as one of my direct reports was when he was able to create a screen name that consisted solely of his first name, which was one of the most common male first names out there. I was able to help him do this because I saw that the screen name had recently been deleted.

In those days, if a screen name on AOL was deleted, it was kept in a held file until a designated time in the future when long enough has passed for the name to have been forgotten as having been associated with the person who deleted it.

AOL employees had a special tool that allowed you to "reserve" a deleted or unused screen name for a certain person.

Getting a choice screen name was a pretty important thing; heck, at the time, even Steve Case hadn't gotten screen name "Steve." (His was close to it, but not quite). (One executive had a healthy-enough sense of self-esteem to have grabbed the screen name "God." I'm not saying who, though).

In fact, certain screen names were so coveted, we used to hold "Screen Name Auctions" in the Auditorium, where members paid money to buy the right to use a certain screen name. (Usually it was one that had recently been deleted, hence would be becoming available to the general public soon). Members were known to pay up to $200 to buy the right to use these high-status online monikers.

The screen names bought were usually playful, like "LoveBunny" or "CoolChick," or show-biz related, like "Batman" or "Elvis."

The Talk Show

It was the heyday of talk shows hosted by the likes of Phil Donahue and the controversial Morton Downey, Jr. We wanted to create our own online talk show personality, and pioneer our own version of the talk show as cyber-event.

There were a number of chat rooms where 'Linkers could discuss serious issues, and several Auditorium events were hosted by issues-oriented personalities Quantum had recruited, such as Lee Chariamonte of the Ms. Foundation. So this would be a natural extension of those, only it would be held in the more controlled environment of the Auditorium.

The People Connection producers sat down together and created our own character, Max Webbe, who was to be a straight-talking cigar smoking guy who cut right to the heart of issues. He was even depicted in one "Update" magazine.

Role-playing Max was among the first forays I would make into the "big leagues" of cyber-entertainment: the Auditorium. And it was via Max that I would meet the biggest celebrities that were online at the time.

Chapter Five: Zen and the
Art of Cyber-Emceeing: Celebrities Online

My experience with celebrities revolved around my duties with AOL's "auditoriums." These "auditoriums" are big virtual rooms where members can ask questions of featured guests and see their real-time responses. It's a more controlled environment than regular chat rooms, because events are "emceed" and the guests and hosts have special tools to "talk backstage" and check questions and comments sent in by the attendants.

"Jacking In": Dr. Timothy Leary

Dr. Timothy Leary stands out in my mind as the ultimate early-adapter cyber-celebrity. He was talking about "cyberspace" before most people had heard the term or knew what it was. When I joined AOL in 1988, he had already done several auditorium events for Q-Link, one of the biggest celebrities they had online at the time.

Later, he even planned to stage his death on the Internet.

My own experience with Leary online may have been the closest I ever came to getting fired at AOL. Dr. Leary was, of course, the outspoken professor known for his iconoclastic beliefs and behavior during the sixties. He had made many outrageous statements and held controversial opinions over the years. For example, one woman I know who attended one of Leary's seminars said he showed a slide of a something that looked like a big pink Georgia O-Keefe-painting. He told the group it was actually an extreme close-up of his wife's genitals.

Leary could be extreme, and had a past checkered with drug use, but in these online auditorium events we were strictly forbidden to touch on certain subjects, as guests often stipulate for offline tv talk shows.

He was a special guest at one of the first auditorium events I ever emceed. Leary was well-known for using the term "jacking in." This was referring to the "data jack" of the then not-too-distant future in which all your media service — the phone, the television, the online service — would be piped in via a single outlet.

The online conversation during the event inevitably swung around to Leary's "jacking in" concept. Since I was playing the part of fictional and edgy cyber-talk-show personality "Max Webbe," as the event's "emcee," I couldn't resist asking Leary to explain the "jacking in" term, kidding him about its similarity to a certain nickname for masturbation.

Well, it turned out Steve Case himself was in the audience that night (he's very much a user of the product and "trolls" the service regularly). My boss the next morning asked me about the incident, at which point I exhibited so much guilt and paranoia about whether or not I would be fired, that she laughed. No, Steve hadn't demanded I be fired, she said, but he had reportedly noticed the comment and wondered about it.

Perhaps the best way to understand how the event went is to look at an excerpt of the auditorium transcript of that night, dated 2/22/89. (Interestingly, although we were instructed not to bring up his association with drugs, he brought it up himself). [Note: "Joan III" was my co-emcee that night, helping out by pulling questions and comments out of the "queue." All questions inputted by users during the events went into a holding queue on a first-come first-served basis, or "first-in first out" in computer lingo.]

Max Webbe: I think we all expect you'll be a little outspoken.

DrTimLeary: At age 69 I think I've earned that Max.

MaxWebbe: Dr. Leary, you're perhaps best known by the expression "tune in, turn on, drop out." If there were a single message you could espouse today, what would that be?"

DrTimLeary: Ah yes. In my youth. A favorite expression.

DrTimLeary: Today? Probably tune in, turn on and JACK IN with your untapped capabilities of the human brain.

Max Webbe: Untapped capabilities...just what do you mean by that?

Max Webbe: ::puff:: ::showing white teeth:: I know they say we only use a small % of our brain power...

DrTimLeary: By that I mean...JACKING IN is a term invented by William Gibson. It means that you digitize your thoughts. And zap them through the screen into cyberspace.

Max Webbe: By organizing them...?

DrTimLeary: There is a growing number of people in the country who realize that the future of evolving humans involves creating simulated realities which exist in the matrix, or digital space.

Max Webbe: ::pulling up socks:::

***

As the show went on, we focused on Leary's current work. Leary had some advanced thoughts about cyberspace and what it meant for humans in the future. At that time, the term "cyberspace" was still relatively new.

Max Webbe: Now what do you mean by "cyberspace"?

DrTimLeary: Cyberspace refers to the simulated realities that we create…with our computers…I am now working with several groups who are developing hardware and software for building the realities of the future.

***

At one point, we received a question asking about Leary's "trouble with the law," and using our backstage tools that allowed emcees and the guest to communicate without the audience seeing, I asked if he wanted to take it.
"No, not on the law," he replied to just us emcees. Instead, Leary further described his concept of Cyberia:

DrTimLeary: You will be able to shake hands with someone thousands of miles away in simulated contact. You will be able to embrace, kiss, undress simulated friends. Within 10 years most Americans will be spending more time interacting in Cyberia than they do in the meat machine world.

Max Webbe: Sounds like Club Caribe on Q-Link! [Club Caribe was a game on the Q-Link service, where you could create little cartoon characters – "avatars" – and move them around a cartoon beach world.] Interesting.

Then we took questions from the audience:

Max Webbe: Well, there's a lot I want to ask you, but I want to give our audience a chance too. OK, on to the phones. [Clearly I had been watching too much Phil Donahue, on whom, along with Morton Downey, Jr., I based my interpretation of Max. I wasn't really going to take phone calls; I was "pulling" and viewing questions that members had inputted. The questions were visible only to me until I used AOL's tools to "broadcast" them.].

Question: Could computer games designed to help attain enlightenment perform the functions of religion in the future? In your opinion, of course. Dave (a big fan)

DrTimLeary: Dave, I don't know what any of us mean by "religion." We are now developing cross-ware in which teachers and students will be interacting with authors in educational Cyberias.

***

Interestingly, although Leary had wanted us to steer clear of the topic of drugs entirely, here is where he brought up the subject himself:

Max Webbe: We had a question from the audience wondering how you got into this field.

DrTimLeary: For 40 years I have been working cheerfully to develop methods which give the individual power to become more intelligent. Psychedelic drugs allow the prudent and skilled person to activate the drugs that enter the universe of the brain. An interpersonal computer allows an individual to communicate and transmit the new perspective of the "turned on" brain. The history of America from 1960 to 1990 can be summarized from Psychedelics to Cybernetics.

Max Webbe: So you advocate use of psychedelic drugs...for the skilled person?

DrTimLeary: I advocate nothing in that line.

■■■

OK, so much for me stirring up trouble. Leary clearly was not going to get into anything about his checkered 60s past too deeply, as we can see by his next reply:

Question: Dr., if you could do it all over would you do it the same (referring to the 60s)?

DrTimLeary: What I do is not relevant. We have experienced many ways of culture change in America in the past three decades which are converting us from a society of factory workers and office drones into an information society. Psychedelic drugs and electronic home appliances ...VCRs, computers, cybernetic programs are converting your bedroom or your office into an information - educational - entertainment center.

***

The next topic was one that has had much discussion devoted to it in the years since this online conference: the blurring of the line between cyberspace and the "real" world, and how much time spent online was "healthy." Although we had many in-house jokes about people racking up huge online bills (sometimes to the exclusion of other necessities, such as utilities), nobody seriously considered it an "addiction" is those days. We did kid about joining "12-step" programs for people who were spending most of their waking hours online.

Question: Is it healthy to devote so much time to simulated fantasy and lose contact with the "real" world?

DrTimLeary: Who are you to define "real reality"? The realities are: 1) the universe is made up of quantum bits of information. 2) the human brain operates as a quantum mechanical device. The realities of meat and machine are transitory. You will be able to pilot your way around the world of material and fleshly hardware when you operate at the level of the universe.

Max Webbe: & good typing, Leary ;)
[this was a "backstage" comment to Leary from me].

***

In the end, Dr. Leary waxed ebullient about the potential of cyberspace, which was a good note to end on:

Question: Do you see Cyberia as a manifestation of the group consciousness which I see as a function of the subconscious? Would interaction in Cyberia be a speeded up version of the impact that certain literary or religious figures have had on our conscious minds through the connection of the media, e.g. Sherlock Holmes, Jesus, et al.?

DrTimLeary: My answer is yes. The poets, painters, magicians, storytellers of the past have created myths and legends which have had more influence than nuts and bolts reality. The legends of Christianity, Buddhism, Shakespeare. But now we have the appliances to make these wonderful brain visions shared by the people. Within 10 years we can hope that the most deprived 14-year-old will feel confident and comfortable trading new versions of Biblical or Shakespearean or Disney legends... We are basically QUARKS carrying a billion neurons of information systems in our brain and we are now learning how to use this equipment – to share the loneliness of our neurological fantasies. If this confuses you…good. Read William Gibson. Stay tuned. Be prepared for a wonderful cybernetic experience in the near future.

***

Dr. Leary was a good sport, though, through all my attempts at effecting a 45-year old wisecracking male talk show host. After the event ended, he typed, "Max, my son Zachary says you smoke far too much."

OK, so maybe my cyber-moderating wasn't too bad…by 22-year old standards. Let's just say I was the first Ricki Lake of cyberspace.

And what of Dr. Leary? His plans for the Internet in both life and death were grandiose. It was fitting, since he'd adopted the Internet as his next big cause in the 90s, that he'd always said he was going to commit suicide online. He didn't wind up doing that…but he did do something that might be considered even more radical.

One longtime auditorium emcee recalls Timothy Leary going into detail about having his head frozen after his death — he said that he had an insurance policy that would cover that aspect and that it would cost a small fortune to have his whole body frozen and maintained.

(This emcee also remembers Leary saying "shit" a lot, but backstage expletives notwithstanding, Leary will always be one of the most interesting and memorable guests with whom AOL's stage was ever graced).

"Leary announced at one of his earlier appearances that he'd arranged to have his body frozen upon his death, in the hopes of being restored to life in a 'new and improved' body (perhaps female, to see how the other half lived), but with his brain intact," said longtime veteran AOL auditorium emcee R.J. Scott.

But then Leary announced at a follow-up event that he'd found that prospect to be too expensive, and that he'd arranged to have only his head frozen, said Scott. "By having only his head frozen, he said he'd save about $85,000.00."

Instead, for $4800, his ashes were launched into space "in the first celestial release of human remains," according to The Associated Press. Along with the ashes of a four-year-old Japanese boy, "Star Trek" creator Gene Roddenberry, and 22 others, Leary is traveling in a lipstick-shaped capsule with a commercial satellite that was launched into orbit in March 1997 from the Canary Islands.

(It "would become Leary's 'Ultimate Trip'!" joked Scott).

Here is how Scott, who had most of the earliest cyberchat dealings with Leary, remembers Dr. Leary:

"When Timothy Leary appeared in the auditorium on Q [Q-Link, AOL's former service for the Commodore 64 computer], as well as on AOL, he preferred to sit in the audience and watch himself 'perform' as I conducted the interview.

Julia L. Wilkinson

"How was that possible? Well, I had four computers at that time, plus a Stratus terminal [the Stratus was the mainframe-like "super-mini" computer which AOL used to manage the service in the early days]. I had two Commodore 128s, an Apple (on loan from AOL), and a PC clone.

"Leary's screen name was 'TFLeary' on the commingled side of the service, and 'Tim Leary' on Q-Link, as I recall. I created the name 'Tleary' on one of my Q-Link disks, and a similar name on one of my AOL (not yet known as AOL) disks. There were now three Timothy Leary accounts signed on to the two services: one was him, two were me.

"There were also two of me signed on: 'RJScott' on Q-Link and 'RJScott V' on AOL. 'RJScott' and 'Tleary' were onstage on Q-Link while 'RJScott V' and my version of his name were onstage on AOL. The REAL Tim Leary was signed on as 'TFLeary' watching from the audience on AOL. (Are we confused yet?)

"It's no secret that many of our computer-shy talents prefer to have a ghost-typist type for them while they answer members' questions. So there I was, headphone on, talking voice with Leary, while manning four computers simultaneously on two platforms of the service.

"The trick was to get a question from the queue on one of the platforms and read it to Leary, then broadcast it. As the question was being read by members, say, on the Q-Link side, I'd duplicate the question on the AOL side and broadcast it. Leary was formulating his answer as I was typing it to the Q-Link audience. Then quickly I'd turn to my PC and type the same (or similar) words to the AOL audience. It was difficult, but not terribly so, once I got my momentum going. Adrenaline rushes can work miracles!

"The problem was Dr. Leary. The man wasn't quiet for a second!

"I'd ask him if he'd slow down a bit, explaining the fourfold task I was trying to accomplish. He was very kind. He accommodated me — only to start asking how my mother was, how the weather was on the East coast, how long I was with the service, etc. He'd also be talking with his wife and his son, Zach, at the same time.

"He had quite a sense of humor too! Occasionally a question would pop up that he chose not to answer. He'd have me in stitches on the phone answering the question for ME while members in the audience were dying to know the answer only to find us moving on to the next question!

"The "Leary Events" were tricky, to say the least, and never without excitement. One never knew what to expect from that man. He was a hoot!

"Leary, I learned, controversial as he was, was an interesting, kindly person in all my dealings with him. There's only one thing members might have enjoyed more than his interesting appearances online: [that is to know] him as I did, as a voice on the other end of the line, three thousand miles away in Beverly Hills, California. His voice spoke softly, yet firmly, with authority. That voice was silenced when Leary died of prostate cancer in 1996. We never met in person. That was my loss."

Dr. Leary — wherever you are out there in the stars — Max Webbe, who is really a 35-year-old woman today, says hello.

***

Wally 'n' The Beav

Back in those early days, we had to sign up celebrities using our charm and wiles, because we still didn't have the name recognition that Compuserve and Prodigy had (or, more likely, the celebrity in question hadn't heard of any online service). Nor did we have huge piles of money to throw at them.

Usually we got people who had been out of the limelight for a while, and intrigued about getting back in through this futuristic new back door, or just people who were curious. I even joked at one staff meeting that we might as well go to Atlantic City to recruit a bunch of has-beens all in one fell swoop.

I remember my then-boss, "Ellen," striding exultantly into my office and saying, "Guess who I just talked to?"

"Who?" I asked.

"Tony Dow!" she exclaimed.

"Tony Dow? You mean The Beav...I mean, Wally?" I said.

Sure enough, The Beav's bro was alive and well and living in Florida...and willing to give cyberspace a shot. In those days that was quite a coup for our small, nerd-centric stable of stars.

Cyber-Comedy

Dilbert Meets Cyberspace

By the time I stopped working full-time, I would have described myself as a middle-manager — the kind that Scott Adams makes fun of in his "Dilbert" comic strips. In one typical strip, Dilbert's manager, who is always depicted as bumbling and incompetent, says he wants to "roll up his shirtsleeves and pitch in" on the work with which his team is inundated. But first he pauses and says, "how do these buttons work again?"

That was how I felt at the end of my management career at AOL, because when I had been a producer, I was used to seeing tangible results of my efforts. As a manager, I had to sit back and let other, younger minds and more nimble fingers crank out forms and create online areas.

Although I was never way up there on the totem pole at AOL by anyone's standards, because of my role as a producer, and the company's autonomous

53

corporate culture, I did get a chance to deal with a wide variety of interesting organizations and notable people. In fact, speaking of Dilbert, I spoke to Scott Adams briefly when we were negotiating to get an electronic version of the comic strip put on the network.

We talked about his decision to move out to California, which he chronicles in one of his books. He was caught in a snowstorm one day, and his old jalopy crapped out on him. He had to walk a long way through the snow and ice, and he vowed if he made it through that experience, he would pick up immediately and move to California. He did make it through the blizzard, and he did move, as soon as he possibly could.

"Dilbert" was not as huge then as it was to become...this was 1992 or '93. But it was one of those things that I had a gut feeling would work. I don't remember where the deal originated, but I championed it in-house, because having read some Dilberts, I thought it would appeal to the average AOL user. (Not to mention that he could have been writing about AOL's own office environment, the resemblance to Dilbert's experiences and ours was so uncanny).

When I was working there, Scott Adams would hang out on AOL. In fact, he told me that since he quit his day job at Pacific Bell, which was so helpful in giving him seeds of ideas, he got some ideas for his strip from e-mail and the message boards on AOL.

***

In the early days, AOL may have had to content itself with lesser-known celebrities, or just the ones who "got it" early, but as AOL ramped up into the mid-90's, all that was changing.

One reason was the hiring of a brash and flashy new executive: Ted Leonsis. From my perspective as an employee, after Ted, nothing was ever the same. We were on our way.

Ted Leonsis: Hollywood Meets Cyberspace

A Decent Proposal

The first time AOL employees officially met Ted, he addressed the company at an "all-hands" meeting shortly after he started as AOL President. He said he'd recently accompanied his son to his first day of school, and in many ways he felt like speaking at that meeting was *his* first day at school. It was 1994, and the company had fewer than 500 employees. (By July 1999, it would grow to over 14,000 employees, from 700,000 to 17 million members, from $90 million to $6 billion in revenue and from $400 million to $120 billion in market capitalization. The stock has split six times, leading to Leonsis's net worth of $500 to $900 million dollars).

But it didn't take "Ted the new kid in school" long to be the popular guy in the class. He squinted out at the audience. "Is Jan Brandt here?" he asked, referring to the marketing vice-president who'd overseen AOL's recent disk-spamming of the United States.

"Here," she said.

"I think I'm going to love you," he exclaimed, citing his admiration for her achievement in the high response rate AOL had generated with its marketing programs.

He then challenged her to meet an even higher marketing goal, and she enthusiastically responded, "I accept," to the unanimous laughter of AOL employees.

The other employees were falling in love with Ted, too. He gave good demo, and even better speeches. Ted had developed a product at the company he founded, Redgate, called "2Market." It was an interactive CD-ROM (which was also to have an online connection to AOL) which allowed consumers to browse and shop an electronic mall of sorts. Remember, this was back in the days when computer CD's, when you did see them, were used for games or word processing applications; things like that. And Internet shopping wasn't around back then.

He showed a photo of a man wearing a red sweater. "You can not only see the products and order them from your PC, but you can also see the same sweater in green, in white, in blue..." And with three clicks of the mouse he easily changed the colors of the sweater on the model. That brought a collective "ooohh" from the audience.

*** *** ***

Microsoftophobia

Through Ted, our company meetings became not so much dissemination of information as entertaining dog-and-pony shows. At one famous "Brand" [as in "AOL Brand"] gathering, around the time when Microsoft was gearing up to roll out The Microsoft Network/MSN as an online service, Ted booked a local Sheraton hotel ballroom and rolled a clip from "Jurassic Park" to symbolize the awesome power of Microsoft and Bill Gates.

Fog rolled in; there were loud sound effects. The whole company was whipped up into a frenzy for months. At that meeting, Ted told us that he kidded us not – Microsoft was an awesome and formidable company. I'd heard Steve Case call them a great company before, and knew he had respect for them. But, Ted said, that didn't mean we didn't want to kick their collective butt in the online services market.

Ted had a bunch of t-shirts printed up that said "Online Services will be Microsoft's Vietnam", and handed them out at the meeting.

He had a giant dinosaur sign produced and had every employee sign it in a show of solidarity. There were green hats with a dinosaur face encircled in red and with a diagonal slash through it, as in a "no dinosaurs" sign. This was actually a double entendre, since "Wired Magazine" had recently come out with an issue declaring AOL a "dinosaur" because of the advent of the World Wide Web. ("Wired magazine…" Ted had commented ironically, "A petroleum-based product!").

He may have been warning us of Microsoft's aggressive track record and immense resources, but he didn't want us to think they couldn't be beaten. Because they had been before; they were not invincible. Intuit had proven that.

Intuit was the company who made Quicken, the most popular personal financial accounting software. Microsoft tried to duplicate their success with a product called "Microsoft Money," but it was far from the hit Quicken was. Finally they tried their Plan B: if you can't beat 'em, buy 'em…but the Justice Department smelled anti-trust and blocked the deal.

Another factor against Microsoft producing a viable online service was that being a software conglomerate could damage their credibility as a free and fair communication medium. Who was to say that if a question was asked in a computer & software forum that they wouldn't get an answer that was biased toward Microsoft's products?

But I think the most satisfying thing I heard was a quote by Bill Gates himself, who said something like if you look in any trashcan, you'll see one or more AOL disks. In my mind, if you can get Bill Gates to say something spiteful about your company, you're doing something right. We're talking about a man so competitive that when his mother brought the young Bill to a psychiatrist, after meeting with him, the shrink allegedly told her not to bother competing with him.

(Another anecdote was that upon negotiations to agree to use the Microsoft Internet Explorer as the default web browser on the AOL service, the head of AOL product marketing evidently made Bill Gates sign a pledge admitting that AOL had built the best consumer online experience in the business.)

In fact, Steve Case was quoted as saying "Fear of Microsoft has helped this company enormously. Despite all out growth and the rise in our stock price, no one here is congratulating themselves. We're totally focused on the future." In the pre-MSN launch days, AOLers were whipped into a religious fervor that manifested itself in crazy work hours and total dedication to their jobs. They all wanted to be a part of the David that might slay the Microsoft Goliath.

"If you're the prettiest girl at the prom and someone better-looking and richer and more powerful walks in, of course you're going to react," said an analyst with Paul Kagan Associates, Bishop Cheen, in Advertising Age. "From the moment everyone knew Windows 95 would have an embedded online service, AOL has been working double-time to expand its subscriber base, its content alliances and its verve."

When The Microsoft Network finally launched, and then quickly downgraded its service to a web-based product, it was anticlimactic, to say the least. But there were plenty of other competitors out there, from MCI to giant AT&T to CompuServe, and of course we have not heard the last from Microsoft.

Jerry Seinfeld: The Enemy

After the Microsoft Network made its less-than-intimidating debut, Ted was fond of saying that Bill Gates was no longer the enemy: Jerry Seinfeld was. He'd talk about how he'd rush home in his car many a Thursday night to see "Seinfeld" on TV. If something on AOL could compel people like that, we'd know we'd won the battle.

It was, in fact, a battle for the mindshare of the American public; for that precious stretch of leisure time after dinner and before bed when "every night a decision is made. I've had dinner, I've talked to my wife, and now what do I do?" Ted asked. "Do I watch 'Seinfeld,' or do I turn on America Online?"

Ted's mission was for AOL to ultimately steal people away from Seinfeld's "must-see" half hour – which would also mean we were stealing them away that much more easily when such compelling TV was not competing with us for attention.

(Later that year, Ted was to say that Jerry Seinfeld was no longer the enemy. "People with lives are," he joked.)

Julia L. Wilkinson

Going Hollywood

Ted Leonsis: how to describe him? The words that come to mind are flashy, flamboyant...Hollywood. Indeed, he seemed enamored of Hollywood and had many connections there. Steve Case said "he's a showman," and has been known to call him the "P.T. Barnum of the Internet."

Ted brought the television metaphor to AOL, and there was action backing up all that talk. The erstwhile "departments" such as News, Travel and Sports became "channels"; and online areas became "shows." If shows didn't do well, shows would be canceled, just like on TV, not left to languish on the online vine, as some might have at that time. Before Ted came along, our modus operandi was to launch online areas, and then leave 'em up, unless they were clear dogs. Ted encouraged us to be more discriminating when it came to analyzing an online area.

Television was a free medium for consumers, thanks to good ol' capitalistic advertising, and AOL could hope to reduce or eliminate its subscriber fees by using similar methods. This was initially a big change for AOL, since it had originally disdained the ads on the Prodigy service as too intrusive. (Indeed, when I first joined AOL, its policy was to not carry online advertising other than for copy written for its own areas and events on the service. But as it became clear both that advertising could be informational for consumers, and that it didn't have to be so intrusive (as in a small banner ad taking up just a small portion of the screen), AOL's stance toward advertising changed.)

Television had other lessons to teach AOL, as well. One of Ted's friends was the television executive extraordinaire Brandon Tartikoff, now deceased, the man who gave America the smash NBC hits such as "Cosby" and "Family Ties" in the 1980s. Ted brought Brandon to AOL to talk about what he'd learned from his time as a programmer for television, so we might better learn what worked as programmers for online.

It was late Spring of 1996. A group of us involved in the content creation and management at AOL gathered in a mid-sized conference room at the Sheraton down the street. Tartikoff stood at the podium and spoke at length about the things he'd learned over the years in the television programming world. He said that at a recent television programming conference, he thought he was having flashbacks to the past. "The same stars were there that were there in the eighties...Bill Cosby, Michael J. Fox." Hollywood is scared to take a chance on new faces, he said.

Tartikoff felt he was ready for new challenges, and so started his own production company where he could get back to creating exciting projects, including online endeavors.

Tartikoff had an instinctive feel for what would work on tv. (Although, in his modest manner, he made many jokes about one of his rare failures...a bizarre sitcom featuring a monkey). In one legendary anecdote, Tartikoff had a brainstorm for a new show concept that he had summarized in two words on a cocktail napkin: "MTV Cops." The napkin was handed to an influential person in the tv biz, and not much later, the hip show "Miami Vice" was created – complete with its flashy backdrops and interwoven rock music soundtrack.

He said when people pitched an idea to him, he'd say, "tell me about the characters." He wanted to know about each person in the show and what their whole social dynamic was; getting this right was what made for great television. Now he was ready to translate that thinking into the online medium.

Student of Brands

Ted used to say at company meetings that he was a "student of brands." And he was. He would hold employees rapt with his stories about how household-name brands were made. "We can be Nike," he would say. "We can be like Coca-Cola." Attending these meetings was not unlike how I imagine it would be to take part in a religious cult rally. "We are the number-one company in the number-one growing industry in the world."

He'd regale us with tales of brands like Evian. "Evian…this is bottled water for crying out loud! We're talking two guys in New Jersey filling bottles from the tap! And did you ever notice how 'Evian' is 'naive' spelled backwards?"

One of his favorite media success stories was Nickelodeon. "The Disney channel blew it when they decided to have a premium cable channel that would cost consumers more," he said. "They could have had *the* children's cable franchise, but guess who they lost that to? Nickelodeon."

He was fond of asking questions with surprise answers. "What is the most profitable television show on the air today?" he'd query a meeting of producers, all of them as riveted as a pair of Levi's. "Friends" and "Seinfeld" were thrown out as answers. "It's Talk Soup," he'd say with a mischievous gleam in his green eyes. (Talk Soup, for the uninitiated, is the show on the cable station "E!" which re-hashes the most egregious moments from each t.v. talk show over the past week). Then he'd explain how Talk Soup made money by broadcasting the same show many times over the span of a week.

He was also fond of making bold entrances. I remember one gathering where AOL had booked a room too small for the teeming masses of employees who wanted to attend. (This was in 1996, when AOL had clearly outgrown the space it was inhabiting, and had added thousands of employees in a mere couple years' time). We sat squashed end-to-end in that Sheraton hotel meeting room, sweat glistening on our brows. Ted strutted right up to the microphone, paused for effect, and exclaimed with perfect comedic timing, "You guys stink."

It was the kind of pull-no-punches frank thing he was prone to saying. The New York Times Magazine ran a feature on Ted which had more examples of his bluntness: At one of AOL's Greenhouse "pitch" meetings, where would-be purveyors of cyber-content came to convince AOL execs they had a winning idea for an online area, a couple guys came in to sell their idea for a surfing area online. Presumably dressed to impress, they came in wearing coats and ties. Ted's comment was "these are the fuckin' surfers?"

At a more recent gathering of AOL employees, at the Westfields Conference Center in Chantilly, closer to AOL's current headquarters in Dulles, Leonsis tells the employees that we're in the Bronze Age of media development.

"Knock knock," goes his joke.

"Who's there?" the crowd responds.

"Amy Fisher."

When the crowd responds "Amy Fish-"

"Bang!" says Ted. His point is that you're dead before you know it in the online business.

***

Ted was like the Forrest Gump of the computer industry: always in the right place at the right time. According to his bio on AOL, his list of credits include: working on the team that took Wang into the office automation business; working on the team that took IBM into the personal computer industry; working on Apple's Macintosh; and helping lead companies such as BellSouth, Fidelity, EDS, Prudential, Apple Computer, Sun Microsystems, Marriott, Hewlett Packard, US West, Dow Jones, and others into the new media industry.

During his career, Ted launched six personal computer magazines, the "first all-digital multimedia service, called Interactive Information Networks"; and "En Passant," a digital CD-ROM shopping program which grew into a stand-alone company. He's the author of three books, including "Blue Magic," an insider's view of the launch of the IBM PC.

Ted also started his multimedia company, Redgate Communications, in the 80s. Why the name Redgate? According to a story he told, as a student at Georgetown University, he was golfing at a country club with a well-heeled classmate. The name of the club was "Redgate." Ted took one look at the sign and thought it sounded high-class. "When I found my own company," he thought to himself, "that's what I'm going to name it."

D.C. stockbroker Joe Coffey had been a Georgetown classmate of Ted's (although, as Coffey joked, Ted had been at the head of the class and Joe was closer to the other end). He mentioned that in a class picture of a younger Ted, he was quite a bit leaner than the stout man familiar to AOLers.

But Ted took his corpulence in stride. He was, in fact, known to butt a fellow executive "sumo-style." With his dark green eyes and olive skin, he would be a striking presence at any weight.

Humble Beginnings

Ted's father had been a waiter, and his mom a secretary "whose best year was $28,000," said Ted in a Washington Post interview. Ted learned to love sports while playing basketball and street hockey with friends in his childhood neighborhood in Brooklyn, and one of his most treasured childhood memories was a trip to Yankee Stadium. "I had never seen grass that green or dirt that brown," he said in the Post.

But his folks were concerned about the rough neighborhood, and moved the family to Lowell, Mass., where they had lived previously. "My destiny was to bag groceries at a Demoulas Market, move up to cashier, then manager of the produce department, then, if I worked really hard, I would get to manage a store," he said. Other jobs he had included operating a forklift and working in a dress factory.

Although one of Ted's teachers said he didn't think Ted was "college material," Ted did manage to get himself into Georgetown University. He held various jobs to support himself, including working in the office of then-Rep. Paul Tsongas. Ted graduated from Georgetown in 1977, and got a job in public relations at Wang Labs.

He left Wang in 1981 to start his own publishing company, selling it in 1983. But two years later, he bought it back and launched Redgate, "a small digital marketing enterprise that put catalogues on CD-ROMs," according to the Post.

Then one day in 1993, Ted had breakfast with Steve Case, a meeting which ultimately led to AOL's acquisition of both Redgate and Ted. The meeting was suggested by Dan Case, Steve's brother, the investment banker, who had had dealings with Ted and his company before.

One of Ted's favorite stories is about starting at AOL. He asked Steve Case how he could get office furniture. Steve's reply? "Work late." Now when people ask Ted how to get a good parking space at AOL's often crowded headquarters, he replies, "Get in early."

AOL a Lifestyle

There were signs that AOL started succeeding in the mindshare business by early 1996. AOL had increased subscribers 33.3% to 6 million during the year. "We've made AOL become a lifestyle," Ted was quoted as saying in a June 96 issue of Advertising Age. "An indication is that we're selling a lot of AOL-branded merchandise. Not many people buy IBM T-shirts."

Ted rejected the conventional wisdom of the time that said the World Wide Web would cannibalize online business and multiple price options can segment an online company's market.

***

In 1999, having already started and sold his own multimedia company, and firmly established in a leadership position at AOL, Ted was ready to try something new yet again. He, multimillionaire venture capitalist Jonathan Ledecky, and "old hockey hand" Dick Patrick bought the Washington Capitals hockey team for a little more than $200 million. Leonsis owns 65 percent to the other men's 30% and 5%, respectively.

■■■

Owning a sports team was just one of the many items on "The List" of Ted's. "The List" is Ted's "101 Things to Do" before he dies, which he made after a scary plane trip from Miami to Atlanta, when the landing gear failed and the pilot told everyone that in 30 minutes the plane would be forced to make an emergency landing. "I didn't want to die," he said. "I told myself that if I live, I'm going to play offense."

Many things on The List have to do with sports. (Number 12 is Catch a Foul Ball, and he completed that one when he caught one off the bat of Mookie Wilson at the New York Mets training camp). Other items include: Do Stand-Up Comedy Routine at a Comedy Club and Have a Great-Grandchild. Three more sports-related items: No. 65, Own, or Be a Partner in, a Sports Franchise. No. 17: Win an NBA Championship. No. 19: Win a Stanley Cup championship.

Ted's role at AOL has changed over the years, as has AOL. First, he was head of the AOL Brand online service, then he became head of AOL Studios, which is what the "Greenhouse" department evolved into. But, as the Post reported, when it became clear that companies were lining up to have their content on AOL, he moved on to oversee other areas, such as the ICQ instant messaging service, and the MovieFone service.

***

One of my favorite Ted anecdotes was his encounter on a plane with a major executive of a large media company who shall remain nameless. This media company had left AOL the year before with much fanfare. Now, the executive asked Ted about pointing to one of their websites from the AOL service, an action which would substantially increase the site's traffic. Ted let the guy know it was going to cost him.

"Let me get this straight," the executive answered. "Last year, you were paying us to house our content. Now, if we want want you to point to us, it's going to cost us millions!"

***

More and more, people were making the choice to sign on to AOL. In fact, when another big-name executive, Bob Pittman, joined AOL in late 1996, he was fond of telling us how his thirteen-year-old son would go straight to his computer and sign on to AOL after he got home from school.

Bob Pittman: The Energizer Bunny

Bob Pittman was reportedly wooed by AOL management for months. From Case's and Pittman's first meeting over the Century 21 AOL deal (an online area known as Century 21 Communities, and no, there are no gold jacket icons anywhere), a connection was formed between the two men. Both were relatively young and decidedly rich, and they were also both collectively caught up in more than one media revolution of the twentieth century: Bob had overseen MTV from a tiny start-up into a cable tv giant; Steve had founded the largest Internet Online service company.

Steve said they had a dinner together which went on late, as both men talked excitedly for hours about their visions. "It was like talking to the Energizer Bunny," said Steve.

At his first company meeting (held at George Mason University's Patriot Center – we had outgrown any smaller, nearby venues), Bob impressed the AOL employees by speaking with enthusiam and authority, and taking on red-hot questions from staffers. At every AOL company meeting, there was a chance to ask questions of Steve and the top executives at the end.

One bold young programmer got up and queried, "will we ever have an adult area on AOL?" (The "adult area" was a hot-button issue which had reared its head with many incarnations over the years, but just never would die). With a comedian's timing, Steve turned to Bob and asked, "Bob, why don't you take this one?" But Steve took the question anyway, giving a brilliantly noncommital answer. However, Bob did pipe in, saying, in essence, the problem was not new; that t.v. and cable had found ways to handle it, and so would we.

***

Bob Pittman is a man with intrigue of his own. His career reads like something out of a movie script: a boy-wonder college radio station disc jockey, he went on to found MTV in the eighties, as well as Nickelodeon, Nick at Nite and VH-1. Advertising Age called him "one of the most high-profile media executives of the mid-1980s." In the nineties, as CEO of Time Warner Enterprises, he engineered the acquisition of seven Six Flags amusement parks and boosted their attendance. (Ad Age characterized his effect on them as no less than transforming them into "the only potentially nationally branded rival to Walt Disney Co.'s venues."   He then went on to HFS Corp., which owns Century 21 real estate.

New York Magazine ran a piece on Bob and his then-wife, Sandy Hill Pittman, as "The Couple of the Minute." "I'm awed and appalled by the intensity of their desire to climb," one journalist, who knew the couple since their MTV years, is quoted as saying about them in the article.

"Bob's an ambitious guy, but so am I and most people I know," said Roger Altman, then vice chairman of The Blackstone Group. "Bob is not differentiated by ambition or hunger. Actually, Bob's kind of laid-back."

Some blamed Sandy for putting the vaulting ambition in her husband: "She's the type who's *got* to succeed," said one unnamed colleague in the New York piece. "If her business isn't going well, she goes on to something else with tremendous energy rather than trying to salvage it. It was clear always that she wanted to travel with the right crowd." Some co-workers even formed a "Sandy Pittman Hate Club," wherein they send each other stories about the Pittmans and "foam at the mouth and rip out hair."

But if Pittman became the ultimate New York city sophisticate, he started out with down-to-earth Southern roots. He grew up in Mississipi, reported New York Magazine, "as the quintessential baby boomer, boasting that he had the attention span of a flea. He also had the slick soul of a marketing genius raised on nothing but rock and television," wrote Michael Gross in the New York piece.

The son of a Methodist minister, he was "a scrawny church brat" who'd lost one of his eyes after being thrown from a horse at age six. His brother Tom recalled his ambition: "Bob always wanted to be rich and famous. That was clear...he said so."

He bounced from college to college, and although he never graduated from any of the four he attended, he soon made program director of WNBC radio in New York.   He "parlayed that position into starring roles in the station's television commercials and on a late-night rock TV show."

Bob met his former wife, Sandy, on a cross-country flight in 1978. She'd spotted him across the aisle, and decided to read The New Yorker, "'cause that says a lot about a person." They fell in love at first sight, and spent a night at her parents' house, empty that night. After they got back to New York, Sandy found her apartment filled with roses. They were married the following July.

He Wanted His MTV

In three years he moved from program director to senior vice president of the Warner Amex Satellite Entertainment Company, and he was made head of the fledgling MTV.

MTV went live on August 1, 1981, its first music video ever, appropriately enough, "Video Killed the Radio Star," by "The Buggles." (I was a high school junior at the time and became a devoted disciple of the cable channel.) Although the channel lost $20 million a year in 1982, the following year, its ratings climbed 20%, and by fourth quarter had turned a profit.

From those statistics, it's clear the fledgling cable channel was not an immediate success. In fact, Pittman seemed to work magic in getting an all-day, all-night video channel put together to begin with, much less to get the record companies to not only buy into it, but to come to regard it as an essential marketing tool.

Pittman and the management of MTV didn't even have many videos available to circulate at that point. They had to rely on many British-made videos, because that's where the videos were being made in those days. (And by some cutting-edge, quirky performance-artist types like Devo and Laurie Anderson, whose quirky early videos are MTV icons now).

"We launched in August [of 1981]. We were going to be budgeting October/November; we knew by the end of November that we had to convince the record industry that we could sell records," said Pittman in VH-1's special, "Video Killed the Radio Star."

"If we could, they'd make more music videos, if we couldn't, we were dead."

***

MTV "was Bob Pittman's first major national controversy," according to New York Magazine. There was confusion over who had actually created it. He was both given credit in the media and claimed the credit for himself: "MTV was a pet idea of mine," he told the Daily News in 1986. Gross writes that "in fact, the idea had been floating around for years. The 1977 show Album Tracks, with Pittman as host, featured clips of rockers like Kansas and Meatloaf. And Pittman says that at least one all-music channel was on the air in Georgia before MTV."

But "Pittman became the unquestioned leader of MTV," writes Gross. And he oversaw the overhauling of Nickelodeon, launching Nick at Nite, and created the mellow twin of MTV, the MTV-of-the-over-30 crowd, VH-1.

"All I worry about is winning," he was quoted as saying in The New York Times. It seemed to be paying off; he made runner-up to Time's 1984 Man of the Year, Peter Ueberroth.

But it wasn't enough for him. "I didn't want to turn 60 and be known as Mr. MTV," he said. He got the backing of MCA, Inc., and formed, coincidentally, a company named Quantum Media, Inc. (no relation to the aforementioned Quantum Computer Services) "with a mandate to build and buy companies across the entertainment board." It was Pittman's dream to create a "multifaceted entertainment company for the nineties."

"He wanted to be the youngest, brashest Turk — the Donald Trump — of the media kingdom," wrote Gross. After attempting a deal with Details, magazine founder Annie Flanders commented, "He wanted to use us to fund young talent and give it a chance to be realized. He wanted to be in advertising, publishing, television, and give each of them synergy with the others. He had this vision of being the major company in young media."

He and Sandy lived the high life, accumulating modern art, attending all the right parties, and collecting "enough party-page photos and mentions in gossip columns to line the walls of their Central Park West apartment" and their 9000 square-foot converted barn in Connecticut. They were profiled in GQ, Esquire and HG, to name a few.

But he was also involved in cultural organizations and philanthropies. Bob was on the board of the New York Shakespeare Festival and a director of the One to One foundation, which works with underprivileged children.

One of Quantum's first nationally syndicated shows, an unforgettably confrontational hour called "The Morton Downey Jr. Show" created Pittman's second controversy. Another Quantum show, "The Street," was "a brave attempt at TV-verité" — a cop show that used tough street language.

But Quantum ran into problems due to lack of capital, and takeover attempts against J. Walter Thompson, the NBC radio stations, and a television-station group called TVX failed. The TV division "was shaken when Downey turned up in the papers with a swastika painted on his face. The Street was torn up by a writers' strike."

A regular eschewer of interviews, Pittman was quoted as saying "I have nothing to talk about" in a brief phone conversation with Gross for the New York piece. "I'm in the press too much. It looks awful. It's deadly. I've got nothing going on. If there was a reason, okay, but it's deadly. I'm disturbed my profile is as high as it is. I'm afraid to go out," he said.

"You know what? I'm the luckiest S.O.B ever walked out of Mississippi," he continued. "But it's troubling at a certain point. You'd like to get involved, but charities think you're only in it for P.R."

Somehow it seems unlikely that someone like Bob Pittman will ever have "nothing going on."

***

He did have a reputation for having the Midas touch. According to one post in the Motley Fool area, "during Mr. Pittman's tenure at HFS as head of marketing, I believe the stock went up over 5 fold in a span of about 18 months — 2 years. One of the key reasons for this is that Pittman was a genius at cross-marketing. He got preferred vendors to pay their hotels and real estate franchises for the right to be able to offer their products to them when they had always been able to do that for free in the past. If there are untapped ways of generating revenue for AOL through cross marketing and the like, Mr. Pittman will discover them... He is one reason I am looking to get back in this stock [AOL] because he worked wonders at HFS."

But it seems that Time Warner and Century 21 weren't enough of an affair to distract Bob from his first and longtime love. "He's really looking for his next MTV," an executive close to Pittman was quoted as saying in Advertising Age.

"MTV is a tough act to follow, but so far Bob hasn't found it. Maybe he thinks he's the next Mike Eisner."

Time will tell whether Pittman has found a rival to his affections in the form of AOL.

***

If Pittman had brought rockers to television with MTV, AOL was now bringing rockers into the online medium via that unique interactive experience, the online chat. Here are some of their more memorable Internet exploits.

Cyber-Rockers

Courtney Love

Courtney Love has been on AOL for a while. In fact, after Kurt Cobain's death, she was allegedly even personally posting to a Nirvana message board. In her 1995 auditorium event, she was vintage outrageous Courtney. She opened with:

C L0VE   : Ta da kiss my *** so exciting I'm leaving.

"DeeMony" asked, "Courtney, I think you were really brave to that article with Rolling Stone Magazine. I just want to know how do you handle the stress that you must be under? I think your [sic] awesome."

C L0VE   : Xanax lots.

Not one to get mushy, when "Kinderhor" asked "Courtney - What do you got to say to all the fans that THANK YOU for saving their lives? There are a lot of them, and sometimes I do too," Courtney replied:

C L0VE   : Stop projecting or I'll projectile vomit.

Courtney revealed that she was "at present reading 'The Madness of a Seduced Woman' by Susan Schaeffer — sort of appropriate."

Of course, her event was plastered with the asterisks that black out naughty words. When one fan asked, "ARE YOU LYING OR IS COURTNEY LOVE REALLY COMING ON?" she replied:

C L0VE   : I'm here ****weed

She dispelled myths about her bisexuality by saying the following about women:

C L0VE   : I slept with one once and just found myself going WHERE'S THE BEEF?

I don't know if you can still find Courtney on AOL, however: apparently someone named "Crazy Amy" "went insane" on her account, so she left the service.

Most interesting to me, however, was the revelation behind the meaning of Courtney's band's name, "Hole." Evidently, it does not have the sexual meaning some people think it does. According to Courtney, it was...:

C L0VE   : From MEDEA by Euripides...she screams at Jason her husband whose left her for a Greek version of an Orange county B model/Juliette Hohner type...there's a hole that goes right through me.

Another rocker who was an early online adapter was Herbie Hancock. He is perhaps best remembered for his instrumental hit "Rockit" on MTV (with its bizarre video of appliances going nuts in a house while images of Hancock flash on a tv screen).

Hancock had been "hip" to online for a long time. He was one of Quantum's earliest celebrity members. Herbie was a big proponent of the Internet: "On the net, the playing field is even. One can create and display without a middleman. However, the real stimulus must come from the minds and hearts of those who use it," he said in his AOL Live event of May 10, 1995.

Hancock said he planned on making interactive music on his CD-ROM, "among other things," but that he's "interested in more than music. I'm primarily interested in human beings and learning about how I can make a difference."

When asked how he feels about the slow demise of public radio due to lack of government funding, Herbie said, "We have the Internet, don't we! I'm glad we don't have to depend on radio for exposure to the public. Jazz may be affected but there is a solution on the Internet."

Like some other celebrities, Grammy and Oscar Award winner Hancock uses the Internet for philanthropic means. He created the Hancock Foundation to help expose underprivileged children to the creative power of computing technology.

More Refugees from MTV

The MTV-AOL connection deepened when former "VJ" (video jockey) Adam Curry burst onto the online scene.

I once talked to MTV v.j. Adam Curry about possibly hosting AOL's music area. At the time, he was still doing some v.j.-ing on MTV and hosting his own web site. That was a kick for me because I remembered many nights of watching Curry host the Top 20 Countdown on MTV in the 80s. A few girls would make cracks that Adam, with his long, windswept blonde hair, was fine "if you liked the Farrah Fawcett look," but I thought Adam was one of the few guys that could pull the long hair thing off.

Adam did a couple of auditorium events for AOL. He revealed that he owned a computer ever since he was 12..."a Sinclair Z80 with 1K of memory." In one of his auditorium events, Adam told of his most exciting interview for MTV, which was with Michael Jackson. Adam is 6'5", and Michael had evidently insisted on standing on a box, so he would appear taller than Adam. Michael also had a guy there whose sole job the whole time was to make sure his pants kept shining.

Adam later started his own company, with a web site called "cybersleaze" that served up juicy, edgy morsels of celebrity dish.

Lady Madonna

While rock stars like Love and Hancock embraced the web, others were more reticent of this new medium. When Madonna was asked how she felt about her daughter, Lourdes, surfing the Internet, she said in the December 1996 Redbook: "She's not going online! No! If she wants to talk to people, she can invite them over. The people who like going online the best are people who can't sustain a relationship for more than five minutes."

Letting Their Hair Down

Although AOL was getting more and more celebs and rockers to do online chats, some people wondered what the advantage of a live interactive event was over watching that celebrity interviewed on television, which for most people is still the faster-moving medium.

The answer is that at an online event you have an opportunity to find out what you want to know about that star, not just what some interviewer wants you to know. Celebrities also tend to relax in this environment, safe in the womb of their home or office, so they often let down their hair more than they would on a hot stage with lights in their face and an "on" camera staring at them.

Take this casual repartee between Barry Williams (tv's "Greg Brady") and Chris Knight "Peter Brady"). (Could Peter Brady really be a computer nerd?). "TVLandMC" is the event's emcee:

TVLandMC: Treble199 asks: What are you up to now? Chris, do you still have your software company?

KnghtChris: I'm now in hardware. But I go both ways.

Without these online events, would Brady fans the world over know such intimate trivia as why the Famous Family didn't have a toilet?

TVLandMC: Cirrus579 asks this burning question: Why didn't you guys have a toilet!

KnghtChris: We didn't need one. We were the Bradys.

BarryWlms: It was against the law to show a porcelain toilet.

KnghtChris: and something about it being potentially revolting to our audience…would you have been revolted? Barry would plastic, steel, a bedpan or a used spittoon have been better?

BarryWlms: Bedpan. That's really revolting.

KnghtChris: But we never got sick enough to use one. Sixth year perhaps.

And where else would we see Peter needling Greg so relentlessly about his crush on Marcia?

TVLandMC: NiteNerd asks: Chris and Barry, did you have a crush on any actresses you met during the show?

BarryWlms: Ummm...

KnghtChris: No. Maybe a late crush on Eve. Barry is not being honest.

BarryWlms: Susan Dey… 'cause I liked her costumes and her smile.

KnghtChris: Perhaps he can't see the forest for the trees. What about…

BarryWlms: Ha ha ha… that's because he knows I dug Marcia.

KnghtChris: Mo and those long counseling sessions where I helped you unravel the mysteries of Mo. Where I played Yoda to your Luke Skywalker.

BarryWlms: Chris was right on in Mo Country.

***

Then too, online conferences are so much safer for celebrities; no one can flip your trailer:

TVLandMC: Brodan asks: What's the craziest thing a fan has ever done to meet you?

BarryWlms: Uhh…Several of them turned over a motor-home so that we would come out and say, "hi."

KnghtChris: LOL! I'm still rattled by the motor home incident.

BarryWlms: Like we really were going to come out and say, "Hey, thanks."

***

Peter and Greg aren't the only two Bradys who are jokers. Maureen McCormick did an online conference of her own on AOL:

Question: Maureen, if you think any two of the Brady kids were to fall in love, who do you think it would be? — Kerry, NJ.

MMcCormck: Jan and Marcia.

MMcCormck: Yeah, I dated Barry Williams. I had one date with Desi Arnaz, Jr. One or two. And I *wish* I had had a date with Chris Knight, and I *wish* I had had a date with Bob Reed. But the best date was always Eve!

Even Bobby Brady, Mike Lookinland, got into the online conference business:

Question: Who was your favorite guest star on the show?

71

MLooknland: Gotta be Joe Namath. Not just because I'm a big football fan but because that episode was so much fun. That episode happens to contain my favorite line in the whole series. It's when Cindy says, "Somehow he got the idea you were very sick," and I said, "What did you tell him?" and she said, "I told him you were very, very sick."

***

Free Promotion

But it's not just about taking spontaneous questions and connecting with fans. Another bonus for celebrities using the Internet is the ability to promote their projects. Many celebrities run their own web sites, which allows them to not only plug their current endeavors, but to disseminate information as well. Jackie Martling, a.k.a. "Jackie the Jokeman" on Howard Stern's radio show, runs a site called "JokeLand" at http://www.jackiejokeman.com. "I love having a web page and being able to load it with all kinds of jokes and pictures and stories that I've been slowly dispensing for so many years," Jackie told me in e-mail. "And I get a kick out of the e-mail I get, even though a good part of it is as rude as anything I've ever seen."

Jackie got into the web when he was approached by some people who wanted to sell him a place on the Web. "I finagled them into doing it for free," he says. "Then I got turned on to America Online, which I still use, because I'm used to the e-mail set-up and I'm too stubborn to change. I have a cable modem, and fly through America Online, and my Netscape Gold jolts me through the Internet."

Being online can also bring new experiences to people who are used to "seeing it all." Calvert Deforest (known as "Larry Bud Melman" on the NBC version of Late Night with David Letterman), said his strangest experience online was "when I saw myself during my webcast on EarthCam." DeForest also uses the web to hawk his wares: "I get to sell stuff in my gift shop located on my website at www.calvertdeforest.com."

Access to Fans

But for many celebs, being online is very much about reaching out in cyberspace to touch fans. Shawna Koch, Web site Administrator for Roy Rogers and Dale Evans, said "The Internet has allowed us access to so many of Roy and Dale's fans. If it weren't for this access we wouldn't be able to share the lives of Roy and Dale with people from all over the world. We receive the most wonderful email from the fans. I personally have never seen such a loyal and caring group of people. They truly love Roy and Dale.

"These experiences have given me, a person from a younger generation, a better appreciation for what Roy and Dale stood for and the wonderful lives that they have led. For this I can thank the Internet."

For authors and other creative people, the feedback they get online is the hook. Kevin J. Anderson, internationally best-selling author with over ten million science fiction novels in print, including "Star Wars" and "X-FILES," says "It's interesting to get 'instant feedback' from the fans — in many cases, I receive letters about some of my books before I even know they have appeared in the bookstores. I have also used the broad-based online community for rapid research in esoteric subjects for some of my novels — you can always find somebody online who is an expert in any subject imaginable.

"I can answer fan mail more easily, but I also get a lot more of it because it's easier for the fans to contact me. The weirdest part is when some of the people argue with me and disbelieve that I am who I really am!"

Some fans aren't as adoring or amenable. Deforest (the former "Larry Bud" Melman on Dave Letterman's show) says, "Most people are very nice...but boy...some of them ...well, let's just say I wouldn't take them home for dinner." Mostly they would ask him for tickets to the Late Show, he said.

Easier to Keep Up

Martling uses the web and email to communicate not just with his fans, but with his old buddies as well. "For me, being in show business, the best part of being online is getting contacted by old friends from twenty years ago who have followed my progress. It makes you relive all the crap you've been through. And the immediacy of the renewed relationship is jarring. Where there were once three letters over twenty years, it's six e-mails a week. My first e-mails were with a guy I hadn't seen since 1969 who was communicating with me back and forth from Japan, and it blew my mind."

Doing Something Good

Others use the online venue to promote their good works. Andrew Shue, who played "Billy Campbell" on FOX TV's former hit series "Melrose Place," hosted an area on AOL that supports his organization, "Do Something" (now a web site, at www.dosomething.org). He co-founded "Do Something" in 1993 to "inspire and assist young people of all backgrounds to take problem-solving action in their communities."

"Everyone has some issue or some part of society that they are passionate about. We can't do everything, so you have to focus on what you are passionate about," he said, adding, "I appreciate the fact that not once did anybody call Billy a wimp."

***

Of course, some celebrities claim they'll never get the hang of life in cyberspace. When asked what kind of RAM he had on his hard drive during a Prodigy event, Bob Hope joked, "What the hell are you talking about? I can hardly turn the thing on and off!"

***

With America Online's membership growing and growing, it seems almost everyone has an account on the service. In fact, apparently a certain former White House intern had a profile on the service, before it was deleted on Jan. 22, 1997, when the story about her alleged affair with President Clinton exploded in the media. And what was the personal quote attached to that profile?

"Oh, what a tangled web we weave."

Addendum:

Top Ten Guests in AOL Live:

Who are the top ten guests to appear in AOL Live as of Spring 2000? Here they are, highest to lowest, with the total attendance tally on the left:

1.	Britney Spears	234,000 participants
2.	Madonna	121,000 participants
3.	'NSYNC	103,000 participants
4.	Eminem	96,500 participants
5.	Rosie O'Donnell	67,000 participants
6.	Michael Jackson	64,000 participants
7.	Ricky Martin	44,000 participants
8.	Christina Aguilera	41,400 participants
8.	Barbra Streisand	41,000 participants
9.	Rod Stewart	37,500 participants
10.	Hanson	37,000 participants

Chapter Six: Corporate Culture

It wasn't all about dealing with people virtually at AOL. A big part of working there was about dealing with the real people at the company, who, because of the long hours, and similar young age of many of its employees, became what Douglas Coupland would call your "air family."

These were the people you saw during most of your waking hours, and for some, during many of their sleeping hours as well. (In "Generation X," the novel in which Coupland coined the term, he defines it thusly: "Describes the false sense of community among coworkers in an office environment.").

Whether it was because you needed a strong sense of purpose and identity to hang tough as an underdog in an industry that many dismissed as a passing fad, or because the executives at the top felt they had to parcel out some fun if they were to expect such hard work in return, AOL had a tight corporate culture. (Or maybe it was all the t-shirts they handed out).

At the time, we felt that AOL's culture resembled Microsoft's in its employees' work ethic, stock-centric mentality, and competitive spirit. It resembled Apple Computer Corporation, or more precisely, as some would say, "what Apple used to be," in its employees' enthusiasm and devotion to its product.

In fact, our culture at AOL seemed inexorably tied up with Apple's and Microsoft's. That's part of what led to such strong employee opinions of those companies –- from admiration and loyalty to censure and loathing. "Microsoft is like Communism - their methods look really good on paper (i.e. Microsoft Secrets) yet they never seem to actually work," observed one AOL coder.

"Microsoft is viewed with a mixture of respect and dread/contempt," says another developer. "They are difficult to do business with, and there is also a healthy dose of paranoia in working with them."

"I have tremendous respect for Microsoft and Apple, or at least the Apple that used to be," said Marc Seriff, the technical head of the company in the early years. "I think in a lot of ways we patterned ourselves after these two companies (especially early Apple) in that we had an open culture - not too many rules, and that, especially in the beginning, hired great people, gave them a piece of the action and the freedom to do things."

Venue

AOL's current headquarters is Dulles, Virginia, in a large ziggurat-like building that used to be home to British Aerospace. For a long time, however, the company made its home in the aforementioned unassuming block of corporate offices in Vienna, Virginia, just down Leesburg Pike from Tysons Corner mall.

"Steve wanted to move AOL to the Silicon Valley," at one point, said Jim Kimsey, but Kimsey was firmly entrenched in the D.C. area. That stretch of road between the Washington, D.C. Beltway and Dulles International Airport has become, in fact, something of a "Telecom Valley," a center of Internet-centric and telecommunications companies like Uunet, PSInet and Sprint.

When the building was purchased in Dulles, we learned the name of the road leading to our new headquarters would be "AOL Way." This was so reminiscent an address of another certain computer company – One Microsoft Way – that one AOL manager joked, "Why didn't they just call it Two Microsoft Way?"

The building, with its high-ceilinged atrium, capacious interiors and panoramic vista of the Leesburg countryside, is probably considered a step up by some employees who remember the "early days" of huddling in the basement of the Phone Base building, or the shared office space with SAIC Corp. on 8619 Westwood Center in Vienna.

Steve Case, for one, may have bad flashbacks to the time he had to fling a fire hydrant through the carapaced glass window in the stairwell of the 8619 building. He was alone at work one weekend, when a water main had broken and flooded the stairwell, and he was concerned about the damage the water would do to the offices on the first floor if it leaked out the stairwell door, so he hurled the hydrant to let out the water.

They had to call the employees and leave messages for them not to come in Monday. Most of them thought it meant the company was going out of business.

31.2

"There's this eerie, science-fiction lack of anyone who doesn't look exactly 31.2 on the Campus." – from "Microserfs," by Douglas Coupland.

Though the company may have outgrown its birthplace, one thing about this former upstart organization hasn't changed too much: the employees.

AOL's workforce has a relatively low median age. That fact led more than one worker to comment on the office setting seeming more like a college dorm than a place of work. It was not unusual to see, on some Friday afternoons, a spirited water-gun fight between young producers blowing off steam from creating the latest forms or furiously processing art to be installed on the system, or a group of producers swigging Sam Adams beer out at a picnic table behind the building.

You can glimpse a bit of the sometimes wacky environment by this mock "job description" in the company humor publication, the "Quantum Quirk":

"A great position is available for any goofball to join the Services Department of one of the smallest, most unorganized and 'different' companies in the world. Responsibilities include: Wandering around aimlessly, getting drunk on Friday afternoons, juggling fruit in the halls, spraying caustic chemicals at co-workers, doing handstands at your desk for 'rejuvenation' purposes, working with Purple Gorillas, holding dance parties in your office, playing baseball in the halls, laughing hysterically, inventing new and bizarre noises for distraction purposes. If you are an unmotivated, team-oriented rabble-rouser, fill out the following application and arrange an immediate interview."

(The application includes a checklist of "mental disorders" including "delusions of grandeur, paranoid schizophrenia, catatonia, neurosis, and psychosis," all of which are followed by "if checked, WELCOME TO QUANTUM!!!)

But if the average employee was about 31.2, that meant there had to be quite a few workers in their twenties there.

"Dude, It's on Fire"

"Dude, it's on fire!" screamed "Jim," one of my indirect reports, from the phone in the lobby. Sure enough, the AOL digital camera trained on the company parking lot was capturing the unhappy real-time spectacle of Jim's ancient jalopy engulfed in roiling flames.

Jim fit the low age and little life-experience profile to a tee. He once strolled into the office wearing a "Nine-Inch Dick" t-shirt, and even after he was diplomatically admonished for this sartorial snafu, he showed up a couple weeks later in the same uniform, oblivious.

He'd been having problems with his car, an old coupe that was on its last legs. According to his manager, he was just getting ready to take it to the dealership for service. Apparently he'd waited one day too long. After Jim's frantic call to his boss, "Sven," a group of us gathered at the third-floor conference room window, which overlooked the parking lot where the vehicle was smoldering.

Sven turned to me and said, "You have to feel for him; he's only twenty-two. When I was twenty-two, I had no phone."

It's one thing having no phone. It's another having your car transform your company's parking lot into an impromptu barbecue. Having his vehicular inferno captured for posterity on the AOL parking-lot-cam website didn't help his campaign for maturity any, either.

The parking-lot cam wasn't the only digital window into the world of the company. And while it proved useful on snowy days to see relatively how many cars had braved the icy chunks on Westwood Center Drive, it was usually about as exciting as watching paint dry.

Much more interesting to a couple of guys who worked in production and business development together was the video camera they'd rigged in their office to take photos every fifteen minutes. Their favorite shots were of one particularly attractive administrative assistant who may not have realized that her inadvertent poses in their office were captured real-time on their web page.

Under Pressure

As I mentioned before, AOL was a good, but demanding, place to work. It could be frustrating; it could seem dysfunctional; but it also was a dynamic, exciting place with a lot of fun people near my own age.

There was always plenty to do. Or, to put it another way, because the company had to make the most of its relatively meager resources in the early days, there was usually too much to do. But this made the days pass quickly, and, as our technical guru Marc Seriff said, "I've always found it preferable to twiddling my thumbs all day long."

But even Seriff was often under the gun. Two of his biggest challenges, he said, were "consistently having to do things faster than was reasonable." The best example of this was when he had to create the PC-Link service for Tandy. "A contract was signed to produce the service right about New Year's Eve, and it committed us to build a new product by June using an operating system we knew almost nothing about," he said.

The second example could be described by one word: hypergrowth. AOL experienced variations of this phenomenon at different times throughout its quickly expanding history, but the one that Seriff recalls as the toughest was when AOL hit a major performance roadblock in December of '94. "We were unable to get past about 7000 simultaneous users," he said. "I was sent home to rewrite some big chunks of the system that were causing the problems knowing that the company had to stop growing until I finished. It took about six weeks and set the groundwork for the next growth phase." By June 1997, AOL was to have over 300K simultaneous users.

***

Not everybody got into the AOL groove and thrived on the workload. Some people either weren't motivated enough or self-starters enough to fit into the corporate culture. Starting at AOL was like being thrown into a deep end of a pool and having people see if you could swim.

There were war stories: a longtime AOL employee told me about a guy who allegedly left a meeting running down the hall screaming. He never came back.

One of my managers once made a flippant comment that "AOL has sent more people to the loony bin..." However, I do believe he was speaking metaphorically.

But then there was Kip. He didn't get dragged off by men in white coats to a padded cell, but his office at AOL would do just fine. Kip was a reporter, hired from a news organization, and was used to the pressures of the journalistic world. But for some reason, he just couldn't deal with AOL's tools, which at that time were infamously arcane. (Steve Case once commented at that time on the irony of our claim to the easiest-to-use online interface, when our internal production tools were so esoteric). One woman who used the office adjacent to Kip's told me she could hear him through the wall, emitting little gasps of frustration every thirty seconds.

Vacations offered little respite from the daily grind, because often you would return to a teeming mass of fetid e-mail messages, each one its own little nagging action item.

One employee joked before an extended trip to Europe that he was seriously considering faking his own death over there so he wouldn't have to deal with the workload that would inevitably pile up on his return.

But jokes about stress notwithstanding, by the mid-1990s most employees realized they had a job that many people would sacrifice their firstborn to get. And the stock options didn't hurt, either.

White-House Staff Syndrome

Sometimes it was hard to believe the responsibility we were given as employees of a large interactive communications company. As one AOL programmer so humbly said, it was a shock to learn that "we, a bunch of nobodies, were in charge of the online services industry," calling it 'White House staff syndrome.' " But you could look at it also that it had to be somebody, so it might as well be us. As Ted Leonsis liked to say, recycling that famous quote, "If not us, who? If not now, when?"

All the President's Email

Maybe I had White House staff syndrome because I was actually dealing with the White House staff. Shortly after the time of the '92 election, for example, I was working with one of Clinton's press staffers to set up a post office for AOL members to write the President. But his staff either didn't have the time or inclination to check an email box directly, so I had to go into the mailbox, dump the mail to a file, put the file on a floppy disk, and snail mail the disk to the press office.

For a while after the White House forum and responsibility for maintaining the Clinton mailbox had passed from me to someone new, I still had access to the President's mail. Not that I had time to read any of it.

But I do still treasure my foray into downtown D.C., to the Old Executive Office Building, where one of Clinton's press staffers met me to hand me a copy of the hot-off-the-press budget on a floppy disk, for later posting to the White House forum. "Don't leak this out," he cautioned me. "Nobody's seen it yet."

Wow. Such power. (No, I didn't make any sneaky calls to The Washington Post...I was just thrilled to be part of the whole process. And to get out into the daylight for a couple hours).

Hives, Burst Pipes, and a Drive-By Shooting

"Excuses are like assholes. Everybody's got one, and they all stink."

— anonymous

Aside from shots of tequila, some people dealt with the high-pressure, high-workload world of AOL by taking "mental health days" or coming in later when needed. This may be the reason I received such a curious litany of excuses from employees who worked for me over the years.

Kip's, however, were the most numerous, and I must say, the most bizarre.

It started off mildly enough: he phoned me one morning, obviously distraught. "You know that flu shot I had yesterday? I had a bad reaction to it." OK. That was reasonable.

"I went to the hospital, and I've been up all night," he added. This I found odd, but could still believe.

"My pipes burst," he reported from home shortly thereafter. Evidently, his living room floor had been ruined, along with some of his stereo equipment and other electronics. He was dealing with the plumbers now. Wow, this guy sure has bad luck, I thought. I was still inclined to doubt misfortune of this magnitude, coming on the heels of his medical emergency, but I was determined to be supportive.

"I've been in a drive-by shooting," he gasped into the phone a few weeks later. Apparently someone had shot out his back window on Henry St. in north Old Town, Alexandria. When he stopped at the light, the guy in the car next to his waved a gun at his head. Then Kip peeled out, panicked, and that was when his back windshield shattered. I must say, I would always doubt the veracity of that particular excuse.

Maybe Kip started to think I was a jinx, too, because shortly after that he transferred to another department. I later heard he was...uh, asked to leave the company.

The Cat's Chauffeur

Another employee of mine was a real night owl. Ron had various complicated circumstances that caused him to be late. My all-time favorite excuse was the time when he called and said his fiancée had her cat at his apartment, and they were not allowed to have pets in his building, so he had to drive the feline to his fiancée's mother's house in north Maryland. He would be in after this critical chauffeuring.

Shortly after Ron decided to take another job within the company, but was still awaiting his transfer, he gave up all pretense of making excuses. I received an email from him one morning that unceremoniously declared, "I'm going to be in late this morning. Then, I'm leaving early."

Sex, Lies, and Kinky JPEG files

Creating excuses for coming in late, and even extinguishing exploding cars, were the least of your worries as an AOLer. The online world offers many enticing apples for a cubicle-dweller who grows bored and horny after too many hours positioning icons on forms. Even though AOL doesn't create kinky content of its own, it was possible to get to the kinky stuff on the Internet from AOL. All you had to do was know the name of the Usenet newsgroup. Or website. (The integration of Usenet into the AOL service was very carefully done, however, and the sexually oriented stuff on there was not something a novice would stumble across).

However, several employees at AOL who did know how to get to such content, apparently couldn't resist taking a look at it during business hours. One producer just couldn't keep himself from the online eye-candy jar. "I mean, he was just downloading this huge...well, cunt, for lack of a better word...it filled his entire screen! How could you not notice it?" said the coworker who worked just a few cubicles down from him of his colleague's obliviousness.

Some employees did have shame. One told me of his embarrassment when he was barged in upon by a coworker during a working Saturday. His sheepishness vanished, however, when he remembered that this guy was the one who had *taught* him how to "uudecode" and view the binary files.

Still another AOLer, who had a separate directory reserved on his hard drive for these less-than-buttoned-up images, baffled his coworkers who were getting errors on his machine. Searching for the root of the problem, they found his stash of porno files...a gigabyte worth.

Misdirected Mail

Having a "vanity" screen name, like a single first name, can invite trouble. "As an AOL old-timer, I have a first name screen name, but don't use it except for testing and looking around online," said one AOL developer. "I get a lot of misdirected email from people thinking that I'm somebody else. Some of it innocent, some of it not," he said.

"One particular image showed two women engaged in activities that are most probably illegal, and would seem physically impossible. I sent a reply to the sender, asking "What in God's name was that, and why did you send it to me?!?" To which the sender replied "Oh, sorry, I forgot what you like."

Sex and the Single Modem

A word should be said about AOL's own stance on sexual content. For a short while, Q-Link had a human sexuality forum hosted by experts Howard and Martha Brown. When that service was discontinued, there was no real area dedicated to answering people's questions about sex. We listened to pitches from one sex expert, and many dating service producers, but it wasn't until later in the '90s that AOL began warming up to the idea, as it were.

Because I thought such an area would be valuable (and at least equally important, lucrative), I was a big proponent of such an area. I was fighting an uphill battle, though, because the company was understandably reluctant to dip its toe into these steamy waters. AOL had already been burned by news articles about people who used AOL to transmit child pornography (AOL merely acted as the means of transmission, like a telephone company, but it was still a touchy p.r. issue, because the press tended to jump all over these stories).

AOL has definitely relaxed about the subject, though. Today if you do keyword "sex" on AOL, it will take you to the "Relationships and Sexuality" area, where there is a link to an area about "relating, dating, and mating," an "Erectile Dysfunction" site, and sex advice services such as "Dr. Ruth Online."

However, these services are by no means prurient, and are a much-needed forum for people to get information about a subject that they are often too embarrassed to ask family or medical professionals about face-to-face.

***

Inter-office Interactive Communications

Since we all worked for a company that specialized in interactive communications, and we all used AOL at work, it was only natural that we would use AOL's functionality to communicate within the office. But sometimes the use of "instant messages" to carry on a conversation bordered on the ridiculous.

"Instant Messages," or "Ims," are small pop-up windows sent from one AOL member to another, which contain an upper and lower screen for typing a conversation between the two participants.

Ted Leonsis tells of one time when he and Steve Case worked in adjacent offices. "I was sitting here, Steve was sitting there. I was on the phone doing the AT&T/Apple deal; Steve was on the phone doing Netscape and Microsoft, and we were instant messaging each other." Finally Ted rolled his chair out into the hallway and said, "this is ridiculous!"

But Ted and Steve weren't the only ones who got lazy with instant messages; many employees spoke of keeping up lengthy instant message sessions when the person they were typing to sat next door or only a few offices away.

Instant messages sometimes got people in hot water during demos and conferences, too. Ted speaks of a time "I was giving a demo in front of about two thousand people, and Jonathan Bulkeley [then general manager of the Media department] sent me an IM that said 'I'm pissed!'. Everyone laughed. So I said 'let's go with it,' and typed, 'why'?

After the third message from Bulkeley, Ted said, "Jonathan, I'm giving a demo in front of two thousand people!" And there was silence.

Then Jonathan said, "Hi…and now you see the power of instant messages! I was part of the demo and gag all along!"

Having certain screen names, like obviously female ones, can attract the wrong kinds of instant messages, as well. For a long time I had the screen name "Julia." Once when I was doing a demo of AOL at our booth at the Consumer Electronics Show in Chicago, I received this flaming electronic missive: "What are you wearing?" I tried to explain to the person I was demoing to that I didn't know the person who sent it, but I don't think they bought it.

***

The Cyber-Boondoggles: Conferences and Conventions

One of the earliest industry conventions I attended also presented me with my closest brush with Steve Case.

This was during the Consumer Electronics Show in Chicago, circa 1992 or so. A bunch of AOLers and I were there demoing our new Chicago Online area, which was the first in a series of "Digital City" concepts – forums on AOL dedicated to metropolitan areas. One night after a long day of answering attendee's questions, we headed out to a renowned and raucous Second City restaurant called "Dick's Last Resort." The place had matchbooks with naked women on one side, cutesy waitresses who got up on the tables and danced to "YMCA," and its own line of lingerie.

Its featured libation was a large ale known as the "Big-Ass beer." Who was I to turn it down? And good thing too, because the beer helped me stay calm when Steve, seated next to me, leaned forward and, squinting, asked me to read the chalkboard hanging over the lingerie section of the joint. "What does that say?" he asked. I looked over and strained my eyes.

"Uhhhh..." I could always offer eloquence when talking to a CEO.
"It looks to me like it says 'butt floss,'" he said.
"Yep...I believe that's it," I rejoindered.

We later got into a discussion of whether certain forums should be monitored differently than others. At the time, for example, you could say "fuck" in the Grateful Dead Forum, but not in the Cooking Club. Although I approved of the relaxation of the rules, there was a logical argument I didn't know how to refute. If it was OK for one group on the service, why wasn't it OK for another? "The online hosts don't like the policy," I said.

"Why not?" asked Steve.

"Because it's inconsistent," I said. That's what they had been hammering me with. Steve sat quietly for a moment after I said that and looked thoughtful; he seemed to be carefully considering that information. I think AOL has always been relatively open when it came to the "censorship" question, and under Steve's guidance it organized against the proposed Communications Decency Act, which would have limited features for everyone rather than finding a constructive solution for those who needed it.

At the same time, AOL has justifiably approached controversial topics carefully and seriously. But Steve Case has an open mind. By one account, it was a customer service employee familiar with the gay scene who helped convince Steve to allow the Gay and Lesbian Community Forum (GLCF) have a place on America Online.

AOL was not always open to the idea of such an area. The legend had it that once a very senior level executive of AOL had one day walked in to the Services' vice-president's office and objected to all the "fags" and wackos given a voice across the service.

I later heard one of my bosses describe The GLCF as "the ultimate affinity group," and the area went on to become one of the most popular on the service. Be that as it may, you have to realize what a service this forum was for the gay community...many of whom led double lives and/or treasured their anonymity. And being online was the only real way for them to get it.

Hyper-Growth

One day I drove into work and couldn't find a parking space. It was indicative of the amazing growth in customers and employees we'd experienced in the "boom" years — which I consider to be 1993, 1994, and 1995. For a while we had to lease extra parking space from a building up the street from us. (Those long treks to my car were an unwelcome addition to an already tiring day).

When I started at AOL, we had under 100 employees and around 50,000 members. By the time I left, we had over five thousand, and eight million members.

In the early days of the company, you had to struggle to explain not only what you did, but also where your company ranked in the fledgling online industry. As one AOL Macintosh programmer phrased it, it was like "watching the industry go from 'Online service? What's that?' to 'You know Prodigy? AOL is kind of like that, only better' to 'Wow…YOU work for AOL?!?'"

AOL was not shy about celebrating its membership benchmarks. I remember attending a 75,000 member party, 100,000 member party, 500,000 member party, 1,000,000 member party, and finally a 5,000,000 member party.

We also celebrated our initial public offering (IPO) in early 1992, which was when a lot of people realized sudden large gains in their net worth in the span of one day. One employee remembers holding Steve Case by the elbow while developer Ken Huntsman dumped an ice bucket on his head at the going public party.

But things really seemed to go into overdrive in 1993. Jonathan Bulkeley, who was hired from Time Inc. to head up AOL's new Media Group, recalls the period:

"The day I started was March 22, 1993. That week we announced 250,000 members. And my job was signing up media partners."

Jonathan's first week or so at the company was actually rather eerie; his new job was created as part of a company reorganization which hadn't happened yet, so they kept him hidden away in a temp's office downstairs in the building.

"It was crazy…we built a lot of them [online areas]; we did a lot of deals," said Bulkeley. "The intention was to a) bring content to AOL, but also to b) leverage their marketing channels, their customers. Which ended up working OK, not great; one year it was 20% of registrations."

"For the first time you could turn it on, say 'here are recognized brands that have a value. I can get some of them electronically so there must be a value here too,'" he said.

But, Bulkeley noted, people didn't necessarily use them a lot. "I think they bought the product because of it and then ended up using other things — like having an encyclopedia: people don't use it, but they love to know it's there. So it was really important to have those brands...Omni, Time, the New York Times..." he said.

When the company got too big for Bulkeley's taste, he opted to create AOL from scratch in a foreign market: England. "The great part of doing this is I've seen the movie before. I can come over here and do it here," he said of his time in England.

Like several other long-time employees, Bulkeley felt "AOL Classic" back in the States had gotten too big. "It's a big company. It just didn't seem to be that much fun anymore."

"Coming over here, we've got 55 people, all on one floor. I can look out my office and see everyone from my desk," he said.

Partners' Conferences

Another metric for the growth of AOL over the years was how something called the "partners' conference" changed. The first conference was organized quickly and held in Williamsburg, Va. Only upper-level AOL executives attended, so most production staff were not there. It was at this event that the technology then called "hotlinks" (the ability for the information provider to change an icon and the content it links to on the fly) was introduced. Unfortunately, that particular functionality would take quite a while to become a production reality. But then, the partner conferences had a tendency to be a "dog-and-pony show."

Usually AOL would have one partners' conference in the Washington, D.C. area, and another closer to the West Coast, so partners from that region could more easily attend. In winter 1995, the end-of-the year conference was held at the tony Phoenician resort in Scottsdale, Arizona. I was thrilled to be on the list of attendees.

Since AOL had run out of rooms at the Phoenician, and was trying to economize by pairing staff up where possible, my officemate, "Jill," and I were first paired in a room together at John Gardiner's tennis resort, not far from The Phoenician.

Unfortunately my bad travel karma followed us along: we had no running water the morning of Ted Leonsis's big kickoff speech. Jill eventually managed to elicit a trickle out of the faucet and got her hair washed. How she did that with so little water, I'll never know.

The conferences were an opportunity for AOL's partners to hear AOL's execs expound about every imaginable business topic: the company vision, marketing plans, state-of-the-art royalty and usage reports, and how to optimize their creative programming.

The nights were set up to blow off steam, encourage mingling between partners and AOL staff, and to make people do ridiculous things like take a ride on a jury-rigged "bucking bronco" device while wearing a cowboy hat and bandanna.

***

AOL also attended other industry events, such as the Comdex convention and Apple Developers conferences. Usually less work than usual takes place at these functions, as attendees take advantage of the relative freedom and exciting events planned. But some people truly go AWOL.

An AOL software developer tells of a recent last Apple developer's conference: "One of our group went, but we never saw him the entire time we were there. He managed to avoid over a dozen people in the same hotel for a week. It's still a mystery to us what he did during that time. He has since retired."

Dating at AOL, or: Fishing Off the Company Pier

There are many ways of expressing you are dating someone you work with. But whether it's "fishing off the company pier," "dipping into the company inkwell", or "home-stapling the corporate files," it all means the same thing.

Most people avoid office romances due to the potential for disaster if the affair ends unhappily. Some companies frown on such corporate dalliances, worried that the performance of the employees involved will suffer, or that it will be bad for the their wholesome image.

I worked with several people at AOL who all claimed to live by the strict rule that they did not date the people they worked with. However, if you looked around you at AOL, dating, flirting and even marriages amongst AOLians was taking place rampantly. One could argue that in the rarefied social world of a single person in the nineties, who lived and breathed the office culture, and had little time to even pop a frozen dinner in the microwave at the end of an exhausting day, the workplace was the best place of all to find a mate.

Indeed, one of those people with a "hard and fast rule" about not dating the people she worked with claimed that she only used that as a front to nicely turn down co-workers in whom she was not interested.

In fact, at Apple and Microsoft, whose corporate cultures resembled AOL's, there were high-profile couples to emulate. Take the most famous: Bill Gates and Melinda French Gates. They allegedly met at a company picnic when she was in marketing at Microsoft. In fact, she was involved in the less-than-successful "Bob" product, but it hardly mattered at that point.

Guy Kawasaki of Apple, who met his wife while working there, has written about her in his memoir, "The Macintosh Way." (Guy and his wife, Beth, once posed for a MacWEEK ad: "Only two things really excite my husband," says Beth in the ad copy. Beneath that, it reads, "Guy Kawasaki won't tell us what comes in second, but he admits that MacWEEK ranks at the top of his list.").

I myself succumbed to the seductive lure of the office romance. I might have listened to the gloomy foreboding of AOL's online psychic, Scout Bartlett, who foresaw grave danger on the horizon of that relationship, and nipped it in the bud. If I'd paid more heed to Scout, I would have been more prepared for the end.

Scout gave the first "online psychic" auditorium events in the history of interactive services. One night, I attended his event, thinking it would be a kick to get his predictions for myself. I expected something light and fluffy, like you get in horoscopes: "you will soon undergo a period of change with one of your friendships, and pay special attention to your finances this month." But he surprised me by being ominous. He typed something like, "What I see is not good at all. You may be able to weather the storm clouds up ahead for this relationship if you try very, very hard."

True to his word, the relationship imploded a couple of years down the road.

But not everyone had luck finding a date in the hallowed halls of Westwood Center Drive: wrote one single employee under a pseudonym in the company's early underground newspaper, "The Quantum Quirk," "There are a couple possibilities floating around Quantum, but who wants to go out with a computer head? First of all, they don't make enough money to give this bachelorette a Mercedes, and pocket protectors are so fashionable." She dedicated her article to "Quantum Computer Services, the company that has forced this employee to seek companionship outside the walls of 8619 Westwood Center Drive, Suite 200, due to the lack of compatible fellow employees to socialize with on a romantic level."

***

On one of my very first days at AOL (then Quantum), someone saw fit to introduce me around to the whole floor. In those days you could meet people in content development, marketing, and programming all on the same level of one building.

I was introduced to a programmer, Karl, who was "my type," seemed intelligent and kind, and was conveniently single. I think the first way we got together resulted from my sending him a "Stratus message." The Stratus message was the mainframe terminal's equivalent of America Online's "instant message" of today: it popped up a one-line message on the bottom line of your terminal. Co-workers used it to broadcast messages about the system, or, most commonly and importantly, arranging lunch plans.

For anyone who's been there, trying to keep an office romance secret from your inquiring coworkers is like trying to avoid flies in a suit dipped in honey. I inadvertently made it worse myself, however. One day I was having a problem with my email, where the wrong piece of mail would be pulled up from double-clicking the listed subject. If the subject said "Company meeting," for example, the email itself may have been about a project due date instead.

So I called someone from the Quality Assurance (QA) department to check it out. Unfortunately, the piece of mail she decided to check was one of my outgoing email messages to Karl. It was a cutesy love letter.

That was bad enough. But then I made things worse. I was so taken aback that I just sat there, scrolling up and down through the email, my red cheeks facing the terminal, rendered speechless. I couldn't even think of something to say to break the tension of the situation. And the poor woman from QA, bless her heart, just sat there and acted like it was any other email message.

But in the end, I don't think my relationship with Karl did any damage to my career. One of the reasons was because AOL didn't exactly frown on office romances; there were so many of them. In the early days, the proliferation of dating at the company led to a few employees coining the phrase, "Quantum: a hotbed."

***

So our relationship didn't last, but I knew plenty of others that did.

This brings to mind something one of my single programmer friends said. She had a very close relationship with her male boss, and they'd toyed with the idea of dating, but she nixed it. When someone once asked her if she'd experienced sexual harassment on the job, however, she deadpanned, "Not yet. But I'm going to demand it in my next contract."

Fantasies and flirtations aside, there were plenty of couples. Some of the pairings included a production manager and a producer, a general manager and an events manager, and several developers. I knew a couple husband-and-wife programming teams. One of the early directors of H.R. and one of his employees dated and got hitched. And then there's the most conspicuous AOL couple of all: the CEO, Steve Case, and a woman who had been one of his direct reports, Jean Villanueva.

Jean was V.P. of Corporate Communications when the news of the thing got out. There was an item about it in the Washington Post Style section, but I'd heard it through the grapevine the week before. It definitely blew my mind. I was surprised a person like Steve would allow a superior/direct-report relationship to happen; it seemed too risky at that level.

Steve has always been a very private person, and his wife, the few times she was seen at company picnics and such, was not a flashy sort: her dark blond hair was usually cut short and she kept a low profile. She had been his college sweetheart at Williams.

Jean Villanueva, formerly Jean Wackes, was hired at AOL from Genie, General Electric's erstwhile online service, then one of AOL's major competitors. Classically pretty with straight dark hair and eyes, Jean always reminded me of those girls in high school who are on the top of the social order. She was always confident and articulate.

Several months after news of their relationship broke, it was announced that Jean would be taking a six-month leave of absence. Steve and Jean lay low quite a while. The first time I saw them together in a public AOL function was at the June 1996 Partners' Conference. As mentioned before, AOL has Partner Conferences every six months so the executives can give presentations on upcoming features and so AOL's information providers (the "partners") can network and get to know one another. There are usually two or three nights of organized entertainment.

That summer, Ted Leonsis hired Bill Maher of "Politically Incorrect," along with the Washington satirical singing group "The Capitol Steps," to entertain the attendees. I was told by someone who stayed the whole night that Maher had teased Steve about his new love.

At the last Christmas party I attended, the winter after the Maher incident, Steve and Jean attended as a couple. Dancing on the floor in front of the band at Washington's Union Station, they were grinning from ear to ear and seemed deliriously happy. It was like they were the King and Queen of the AOL prom.

***

Steve and Jean eventually did tie the knot. Although there had been wedding ceremonies performed online over the years, in the rooms of People Connection as well as the auditorium, these nuptials were "offline."

In the second piece to appear about them in The Washington Post's gossip column, "The Reliable Source," this one on Thursday, July 9, 1998, part of the copy read:

"Call it wedding@aol.com. America Online Chairman Steve Case, 39, and Jean Villanueva, 38 – his onetime vice president for communications who left AOL two years ago after romance blossomed – have married.

The Rev. Billy Graham did the honors – in person, not on the Net – at Case's Virginia home last Thursday. They met Graham in 1993 when the evangelist did an online chat with AOL subscribers.

Guests at the small family wedding included all five children by the couple's previous marriages. Both left their spouses and announced they were an item in the spring of 1996, after Villanueva returned from maternity leave. The couple said they had not been involved until they separated.

That fall, Villanueva took a six-month leave 'to address pressing family matters.' She did not return to the company, and now works with several area nonprofit groups.

***

It went on to report that the couple was honeymooning at a media conference in Sun Valley, Idaho. "We hope at least they'll skip a few seminars for some quality time in a hot tub."

I had spoken with Jean a few times over the years, and attended services/marketing meetings where she was present. I wasn't surprised that Steve would fall for her; she seemed a very confident and aggressive professional, and is a very attractive person. Even some of her more knee-jerk reactions, as when dealing with the very sticky and stressful online porn accusations sometimes levied against AOL, like saying if there were inappropriate content or areas on the service, we would "shut it down, and shut it down fast" – couldn't dampen her overall competence.

She could also be very affable and kind. I once ran into her at the spa in the Phoenician hotel, after AOL's lavish partners' conference there in the Winter of 1995. I was voicing concerns about my husband, alone at home with our then-two-year old. "He can handle it," she said, urging me not to succumb to guilt.

***

The below are dating tips based on my experience at AOL, but they might be used for any modern company where hormones are raging.

Exercise 6.1: Go to a "Beer Bash" or other company function. Scope out all members of the opposite sex (or same sex, if that's what you're into), without trying to look too obvious or blowing a neck tendon.

Exercise 6.2: Make lots of lunch dates with anyone who's remotely attractive. Use the time as free testing of the compatibility waters. If you're not interested in the person, order an egg salad sandwich and don't bother trying to eat neatly.

Exercise 6.3: Women: Call the AOL Operations Help and summon a hardware technician to your office. Make cooing sounds about how smart he is to be able to fix your machine so quickly. Men: Get a job in Marketing, where the female-to-male ratio is high.

From the Politically Correct Department

Disclaimer: Some interoffice romances are ill-advised. In one example from another high-tech company, a woman who was stalked by one of her creepier, apparently un-fireable, co-workers.

The Earth Moved: Layoffs

"I've never seen Steve Case nervous before," one coworker of mine said. "Today, he was nervous."

As I got to the doorway of my office, I spotted my boss of the time looking stressed out and smoking in the hallway. She explained to me that she'd had to let several people go as part of the company's layoff. Apparently the company had been too free and easy with spending the money they'd gotten from Apple Corporation. The Board of Directors demanded 15 names from the company to be let go, so the story goes.

You never forget your first layoff. This was late Fall of 1988, barely two months after I started at the company. Fittingly enough, I straggled in at 11 a.m. because I had been online until the wee hours the night before. Back in those days, I "emceed" my own auditorium events and was often up late online.

Later that day, there was a company meeting, where Steve Case got up in the auditorium and explained the numbers showed they had to let people go. Then he gave a slide presentation, where on one slide the actual names of all the people were listed. His voice quavered on that one.

Reorganizations – "Reorgs"

They say that the only things you can count on in life are death and taxes. But if you worked at America Online, it was also pretty darn reliable that there would be a reorganization every eighteen months or so.

After the meeting, we had sessions with a human resources employee who had us get into groups to talk about our feelings vis-a-vis the event. (Which struck me as a little odd, since the company itself was the perpetrator. But still, you had to appreciate their effort). "Rick," one of the guys I had shared a big office with, said "I'm just glad it's not me bent over with a poker up my rear-end, because it easily could have been."

Carol, my first office mate, agreed. "I think we in Services (the content development part of Quantum) feel that all of us could do each other's jobs if we had to switch places."

"Lisa," one of the women who got laid off, was taken out drinking to an Arlington pub by several former colleagues. The next day, one of them reported she'd been holed up in the bathroom there for quite a while. After tying quite a few on, she kept repeating "got no fuckin' JOB" and high-fiving them.

The layoffs weren't as bad as the so-called "aftershocks," those mini, sneaky layoffs where two or three people disappear or "leave to pursue other interests." The disappearance of one such trio was satirically depicted in the company's underground newspaper as "The Quantum Bermuda Triangle," with pictures of three planes, each with a different missing person's name across it, circling below a map of that island.

One senior executive, viewing the image, declared, "That's *cold*."

Layoffs all but disappeared in the boomtown years of 1993-1995. Then came 1996.

No Bozos?

"Do you mean to tell me we of all the people we hired, we have no bozos?" said Ted Leonsis in one meeting I attended. He knew of no one who was fired in the company in a long period of time.

But AOL's fortunes had taken a beating in '96 with Bill Razzouk's departure [Razzouk, former key executive at Federal Express, had been hired shortly before as AOL COO], the perceived threat of the ISPs and flat-rate pricing, the system outage, and AOL's falling stock price. AOL was forced to take a hard look at its business to reach profitability.

It was coming around again. In November 1996, the carnage was severe. All around the company, stressed-out employees were sending each other instant messages, finding out which departments had been hit the worst. Since AOL's latest production tools put a lot of power for form and art creation in the hands of the partners, producers were not as needed anymore, and many of them went the way of the Dodo bird.

In all the personnel changes, however, some administrative i's apparently didn't get dotted. One employee was widely reported to have been on a layoff list. He fielded sympathy e-mails all day long, even though he was still gainfully employed.

Another who was not on the hit list mistakenly received an entire termination package at his home address, according to one developer.

Abrupt Departures

You either got into the AOL corporate culture, or it was determined you were a square peg in AOL's round hole, and you didn't last long.

One story that illustrates this is the time AOL hired "cyberspace's first talent scout," who I will call Jim Burke. A feature article was done on Burke in AOL's then-print publication, "Multimedia Online." Jim was part of AOL's

"Greenhouse" department, Ted Leonsis's baby, which was like a television studio whose mission was to create and seed cyber "shows."

Greenhouse was open to the submission of proposals from the public. Greenhouse staff would weed through thousands of submissions, cull out the best ones, and invite a select few to "pitch" their ideas during a Friday meeting.

One such pitch I attended involved two former TV Guide employees who were proposing a forum for book lovers. Jim made an odd comment during the staff discussion of their presentation, that "you make a choice in your life what you're going to do, and these guys chose to work for TV Guide."

"Actually, Jim, TV Guide is one of the top-selling magazines in the world," piped up one of the Greenhouse's technical producers.

Shortly after that meeting, Jim was fired.

Often the fast pace of events at AOL would make the print reporting of them obsolete. About a month later, AOL's "Multimedia Online" magazine ran a feature article about Burke. By that time, he was long gone. In AOL time, of course.

Bill Razzouk

Bill Razzouk, Chief Operations Officer for only four months, had already closed on a million-plus dollar home in the D.C. suburbs, but the party line was that he decided he didn't want to move his family from Memphis after all.

The real story was that Bill didn't fit into the corporate culture. He reportedly called 7:30 a.m. meetings and didn't let anyone in who was even a minute late. You have to remember, you're talking about a company where a lot of employees considered 10 a.m. starting time (although in fairness, they all left later in the evening as well).

Bill had been quoted in the press lambasting AOL's sloppy business habits. He said he'd tried to phone several high-level executives, and had gotten phone mail for each of them. This, in his mind, was unacceptable. Whether that is acceptable or not, some may have thought that what was unacceptable was a top-level executive feeding negative quotes about the company he'd just been hired by to the press, which was tough enough on AOL as it was.

The moment I knew it wasn't going to work out for Bill was at the June partners' conference the year he was hired. He was to give a speech about the company's call centers and related processes. As he got up to the podium and started to talk, it was apparent his microphone wasn't working. He made a few more attempts, to no avail.

Finally a technician got the device working, and Razzouk cracked into it, "Sort of like trying to connect to America Online." This drew a few startled laughs from the audience of AOL executives and business partners, but mostly hisses and boos.

It wasn't long after that a memo announcing Bill's untimely departure was handed to each employee by a security guard as they headed into the office building one morning.

At one point, when Razzouk had suggested that there be a "formal day" one day of the week, so that clothing like ties and dress slacks would be mandatory at times — a pot-shot at the perpetual casual state of dress AOL employees were in — one executive intimated he would be going a bit far with his new regimentation. That joke may have done more damage to his prospects than anything else, because if it's one thing AOL loves, it's t-shirts.

Recommended Attire

Every day was "casual day" at AOL. The freedom to work in comfortable clothing was not unique to AOL, but common in the high-tech industry; let the programmers wear their t-shirts and jeans, was the thinking, if that helps them crank out better code.

The anathema of the suit was made evident by one employee who dressed up in a coat and tie for Halloween. Another Halloween costume was a woman who dressed as "Dilbert," complete with wire stuck in her tie to make it look like it was sticking up.

Probably the most pervasive items of clothing were t-shirts. Not just any t-shirts; but AOL-made t-shirts. Virtual Places t-shirts, GNN t-shirts (AOL's erstwhile Global Network Navigator), Greenhouse t-shirts, Customer Support t-shirts, Entertainment channel t-shirts, etc. etc. AOL printed up a T-shirt for practically every press release.

Steve Case was even chosen to represent casually dressed employees everywhere when he posed for a "Gap" ad: "You do your best work in khakis," read the ad copy.

But there was one occasion for which Steve and every other AOL employee involved with partners or prospective clients would dress up: the meeting.

The Art of the Cyber-Deal: Meetings

I never enjoyed business meetings. I guess that's not much of a secret because on one of my last days as an AOL employee my then-boss said to me, "you'll never have to go to another meeting," as if I had won the lottery.

The tone of the AOL business meeting changed distinctly over the years. In the early days, because we were one of the smaller online services, we had to be extremely aggressive and use abstruse selling points like being good people to work with and having the easiest online interface.

Later, as online and Internet awareness crept in, more and more companies and media partners began understanding the importance of getting a foothold in this groundbreaking field. Many of them took the mindset that although it would probably not be profitable early on, they would be left flat-footed if they didn't get on the train.

(Microsoft is often held out as one of the companies that did not get on the Internet bandwagon quickly enough. However, with their phenomenal resources, they made up rapidly for lost time. Apparently Bill admonished folks at the company when he learned they were falling behind).

I met with lots of different "information providers," including sources for weather, news feeds, financial services, real estate, gay and lesbian concerns, newspapers such as the Army Times, The Washington Post, and the New York Times; and magazines such as Car and Driver, Business Week, Woman's Day, and Wired.

Early in my career, things were a lot more free-wheeling; like the times I went to comedy clubs, for example, and invited a comedian to do his schtick online in a conference room. As we got into the nineties, AOL separated business development from production, so it was no longer my role to solicit new content.

But in the early days, I sought out and developed several humor services. One was the Comedy Club, which was a collection of message boards and archive libraries. I used to troll the Usenet newsgroup rec.humor and collect their "canonical" (complete) lists of jokes on various topics, such as blonde jokes, O.J. jokes, lawyer jokes, and funny bumper stickers. Then I would edit them and post them to the club library so anyone could download them without having to sift through all the crap in rec.humor like I did.

While I was working with the News area, I thought it would be a good idea for AOL to have its own editorial cartoon, so I called my cousin Signe Wilkinson, editorial cartoonist for the Philadelphia Inquirer, and asked if she knew anyone who might be interested. (A little nepotism never hurt anyone). One of the people she suggested was Mike Keefe, who became AOL's first exclusive original cartoonist. Today, Keefe's dePIXion Studio's runs a whole multimedia area called InToon, complete with "Talking Heads" QuickTime and AVI animation files, cartoons and contests.

As the years went on, I began wondering if I was hexed with some kind of bad meeting karma. My coworkers and I used to joke about it. It started with small enough weirdness: people came up to me and asked odd questions when I was walking through Manhattan. Understandable enough, as I wasn't a native, and New Yorkers can sniff out aliens.

But it got more and more sinister. After one meeting with The New York Times, a subway bomb had gone off and the streets were chaos. My colleague and I waited over an hour to hail a cab. On another day, I was splashed with mud by a passing truck as I headed into a meeting with HarperCollins Publishers. I had to run over to The Gap across the street and buy a new outfit before the meeting.

Going home that evening, we were stalled at LaGuardia for hours since the control tower had received a bomb threat; we were told that it was a rare if not unprecedented event.

At another meeting with HarperCollins, this time in San Francisco, a bird flew militantly into my coworker's hair. (She had a bird phobia). She concluded I was bad travel-luck and avoided further trips with me.

***

The Greenhouse "Pitch"

One of the more entertaining business meetings at AOL was the Hollywood-like experience known as the Greenhouse "pitch" meeting. As aforementioned, the "Greenhouse" was the part of AOL that solicited proposals from dedicated cyber-entrepreneurs and invited the cream of the crop to come to Vienna to "pitch" their idea to the AOL staff.

The typical pitch meeting would begin with the would-be entrepreneurs giving a presentation about the area they wanted to produce online, then taking staffers' questions, and finally, leaving the room so AOL staffers could discuss the merits of their idea and suggested implementation.

The New York Times Magazine described one such pitch, where two Californians took a stab at describing an online surfing nirvana. "Assuming that the members of America Online's staff would be corporate types, Mark Bertignoli and Bob Wolfe wore newly purchased suits and ties," wrote Jesse Kornbluth. After they gave a "buttoned up" presentation, Ted Leonsis was prompted to fume, "These are the [expletive] surfers?" So the two returned for a second presentation later wearing caps, sunglasses, Hawaiian shirts, shorts and sandals, declaring they were "back, bitchin' and bad."

Bertignoli described a surfing forum where they would "beam out from the beach," with live pictures uploaded every two minutes, plus features like trivia contests and the "surf injury of the month."

But the early days of Greenhouse pitches were something else. AOL had invited all its members to submit their best proposal for an online area...and was inundated with all kinds of ideas. So many that it recruited a number of the creative services staffers to pool their resources in rating the documents. We got everything...one guy wanted to post an area of .gif files of various young and attractive models. Another essentially had an idea for recreating a human being as a robot...exactly how he tied that to creating an online area, I can't remember.

Then the areas were winnowed down…only a few could ultimately make it. One that made it to some of the final cuts was a women's forum; another was an area for high schoolers to exchange information and blow off steam.

Several areas from the hundreds of proposals eventually made it to the online world, and after that, only the ones that could sustain AOL members' attention would stay up.

But AOL soon had bigger fish to fry, as it cast its net into the international waters.

The International Business Trip from Hell

More and more AOL old-timers were being recruited into the brave new world of AOL International. It offered exciting opportunities, but they weren't for everyone. (I turned down a job with AOL Canada, for example, since it would require too much time away from my one-year-old. And the lifestyle could be even more frenetic and stressful than it was for an AOLer stateside.)

Bill Gorman, formerly an employee of AOL International, recounts this tale of the "the International business trip from hell":

DAY ZERO

5 PM   Flight leaves from Dulles airport [with Jack Davies, V.P. of AOL International]

DAY ONE

7:30 AM   Arrive Paris DeGaulle

7:30 AM   Picked up by Anthony Khan AOL/Paris, who drives us 10 minutes to a nearby budget motel to take a shower. The showerhead is broken, basically leaving a "hose" that we use to shower with. Race back to DeGaulle.

9 AM   Flight to Milan.

11 AM   Arrive Milan, where my memory is of the huge airport concourse with 40' ceilings, entirely filled with cigarette smoke.

11 AM-4 PM   Cab ride to a meeting with Olivetti, Meeting with Olivetti, cab ride back to the Milan airport

4-6 PM   Flight to Brussels arriving about 6PM

6 PM   Pick up a big Volvo rental car and drive 200kph for 2 hours to Luxembourg

8:30 PM   Dinner with Europe Online principals until 12:30AM.

12:30-2 AM   Meeting with Jack and Anthony.

END DAY ONE

DAY TWO

8 AM-6 PM   Meetings with a variety of communications and media companies.

7 PM-Midnight Banquet Dinner with Europe Online principals and investors

Midnight-2 AM Meeting with Jack and Anthony.

END DAY TWO

DAY THREE

7 AM   Flight to Paris with Anthony [Jack flies to London to attend conference]

9 AM   Arrive Paris and survive Anthony's driving downtown to the AOL/Paris office

3 PM   While working in the AOL Paris office, fatigue overtakes me and I lose the ability to concentrate, although my hotel is only a dozen blocks away I become disoriented and it takes me an hour to walk there.

4 PM   Just as I lay my head on my pillow hoping for some rest, the phone rings, it's Jack, he's calling from London and is waiting for his flight to Paris.

4:30-6 PM   Peaceful sleep as Jack travels from London-Paris

6:30-9:30 PM   Meeting with Jack and principal from Matra-Hachette

9:30-12:30 AM Dinner meeting with Jack and principal from Reuters New Media

END DAY THREE

DAY FOUR

7 AM   Flight to Munich. Although I *always* carry on my luggage, Jack convinces me to check this one time.

9 AM   Arrive Munich with no D-marks. I wait in an endless line to exchange money, just as I reach the window, Jack pages me. He doesn't have money, he was just wondering where I was. Now we have no time to wait in line for money, so we go out to the cab line and find a cab and ask if we can pay with a credit card. The cabbie agrees, and we break the land speed record to arrive on time for our meeting downtown.

10 AM   Our cabbie declines payment on arrival downtown, preferring to *wait* for us to finish our meeting and then drive us back to the airport.

Noon   We dash back to the airport with the same cabbie.

1 PM   Cabbie comes into the terminal with us to "see that we're OK." We tip him very well.

2-3 PM   Flight from Munich-London

6 PM [Eastern Time]   Final flight arrives back at Dulles. My luggage doesn't make it and spends the weekend in Frankfurt.

END DAY FOUR

Four days, seven cities, seven flights, six high-speed car rides. ;)

Big in Japan

But it wasn't just the European theater that AOL International was after. On Gorman's first trip to Japan with Steve Case and John Sculley, recently former Apple CEO, he narrowly avoided what could have been an awkward conversation. "Waiting for a meeting at Mitsui [our ultimate JV partner] to begin, a Mitsui executive said, 'I understand that Mr. Case and Mr. Sculley worked together at PepsiCo some time ago.' I considered my answer carefully, knowing that at the time John had been the CEO, while Steve had been driving around the Midwest testing pizza toppings."

"Yes they did," Gorman replied simply, and left it at that. "Both Steve and John were amused by the story later, for different reasons," he said.

The "There" There: A "Business Trip" in La-La Land

Although my own travels for AOL did not take me across any oceans, I did see many cities in the U.S. But the most memorable trip was to L.A.

Daphne Zuniga of "Melrose Place" was in the doorway, we were all on our second martini, and this day-long boondoggle we called a meeting had degenerated into a botched seduction between representatives of these two large international media companies.

The trip was illustrative of how AOL had changed over the years: now we were concerned with t.v. and movies as much as with print partners; the companies were bigger, the meetings more flamboyant.

Dorothy Parker said of Los Angeles, "There's no *there* there." Our meeting with this television production studio and major media corporation reinforced that observation, as we were whisked from one disconnected spot to the next in a frenzy of activity.

The television production company was affiliated with a certain non-cable broadcast network. It was May of 1996. I flew out of Baltimore-Washington International airport with my coworker, "Catherine."

The production company representatives had sent a limousine to pick us up, so from the moment we got off the plane, this trip was different.

Our first stop? Universal Studios. We had lunch in an outdoor restaurant where my departmental partner "Catherine," the business development reps "Nell" and "Mandy," a couple AOL legal reps, and I met the studio guys. There was Kevin, who owned the company; Adam, who was his lawyer; and Alan, their technical guy.

Although they had faxed us the schedule ahead of time, we must have found it hard to believe that we were really going on Universal's V.I.P. studio tour, because we were all wearing business attire.

But sure enough, right after lunch, we all tromped over to the little golf-cart like vehicles that shuttled groups of people past famous movie and t.v. sets. Our little golf cart chugged up and around the set, passing the original "Bates Motel" from the movie "Psycho," autos from movies and t.v. shows, and Hollywood sets that had been the backdrop of many a feature. I felt silly shuffling around the amusement park in my blue business suit and chunky high-heeled loafers.

At the end of the tour, we realized Ed, "Al Bundy" of Married with Children, had been on the tour with us the whole time, but with his hat, sunglasses and facial hair, nobody recognized him on our trolley. Someone in our group had heard Ed telling his wife about certain episodes of "Married" that were taped at some of the studios we passed.

After the tour, they had dinner planned, and the evening's entertainment: Herb Alpert and the Tijuana Brass at The House of Blues. The House of Blues, for the uninitiated, is a "hip" destination in L.A. that is a bar/restaurant and concert club in one. When it's time for the evening's show, the top restaurant floor swings open to show the stage below. It's like something out of the gym pool scene in "It's a Wonderful Life."

But after the beers we had washed down at the House of Blues, my jet lag started catching up with me. As the second set began, I searched out one of the few chairs I could find and planted myself in it. I tilted my head back involuntarily and fell asleep. I awoke to find Mandy laughing and pointing at me. Nothing like corporate camaraderie. (Not that I can entirely blame them. It was rather pathetic: Here Herb Alpert, who rarely plays out, is cranking out all his songs, and I'm nodding off in the back of the room).

Later that night, as we drove back to the hotel, I announced to my fellow AOLers that I thought I was getting my second wind.

"You let us know when that happens, Julia," said the guy from Legal.

***

The next day, Kevin sent a car to pick us up and take us to their production studio offices. The day was slow to begin, but after everyone got settled and Kevin and Adam had checked their morning email, we had a meeting about the staffing and scope of their proposed online site.

These guys were my first experience with a television outfit creating online content for AOL. Wisely, they didn't just want to do what was at AOL disdainfully called "repurposing content" onto the online service, but they wanted to create original features for the network that would only live on AOL. (This "only on AOL" distinction was becoming more and more important for AOL, since AOL was constantly being expected to justify its differentiation from the increasingly ubiquitous World Wide Web).

Kevin and Adam were bringing the thinking of television, complete with dayparts and targeting content for the differing audiences throughout the day, to their online site. This was very good. However, our task was to bring them to reality as to what would be a realistic scope of content to start out with, and what they would be able to maintain with their existing staff.

The meeting didn't go on too long before breaking up into smaller groups, where Adam and I sat down and went over his more technical questions.

After lunch at one of those quintessentially Californian cuisine restaurants where the entrees consist of tiny, artistic portions, it was off to the Network Z's studio offices. It was in a trailer on a production lot, just like those you imagine stars use on movie sets. The upstairs trailer office of the network executive was plush, however, with nice modern furniture and new carpeting.

▪▪

A handsome man greeted us behind a large, corporate desk. "Mike" was younger than I expected, with dark, gelled hair.

The AOLers lined up on the plush couch; Kevin, Adam and Alan from the production company sat across from us in chairs. Mike sat in the middle behind the desk, facing away from the wall.

After cursory introductions, Mandy endeavored to get a business discussion going. She was seated closest to Mike, in a chair adjacent to our couch. "I'd like to take a couple steps back before we take a step forward and review where we've been," she began.

Mike looked taken aback by all this Eastern professionalism. "Wow," he said. "That sounds like the polka. Or what is it...the hokey-pokey?"

Mandy continued, naming all the wonderful things AOL brought to the table, like proven success in the online market, the largest conglomeration of eyeballs in on online service, etc. (Inherent in the "proven success" comment was the subtle allusion to Network Z's parent company's failure at their own online service venture, despite their deep pockets).

"And what would you say you bring to the table?" she asked Mike.

When Mike then went off on a bit of a tangent, she said, "So you can't...put your hands on exactly what you bring to the table?"

Mike palmed a soft blue toy, one of those "stress relieving" office grabbers. "Well, it's a squishy thing," he said, squeezing the blob as if that was the object to which he was referring, not this intangible deal that hovered in the air like a putative virus. He gave a spiel about their awesome distribution channels, vertical markets of the television shows, blah blah blah.

So nothing was accomplished at that meeting.

Next, it was off to a visit to a movie set somewhere in the "garment district" of LA, where reams of fabric were sold and sweatshop workers labored to make the dresses of the boutiques, or wherever their ultimate destination was. More of the feeling of the "no there there."

The filming was done in a nondescript brownstone, and up a flight of rickety stairs. A plain room was draped with photo deflection canvas and other lighting accessories. When you took in the room as a whole, it didn't look much more professional than a high school stage. We crept into the back of the room, and Kevin nodded hello to the director and whispered to us to try to keep still and quiet.

However, the set itself looked like an ordinary kitchen/dining room, which I guess was all it needed to do. I didn't recognize any of the actors. The actress looked about 25, thin (of course), with long ash blonde hair, pulled back conservatively with a headband. The actor was about the same age, nice-looking but not spectacular. He could have been a character on any soap opera.

On every take, their lines changed a little; it seemed like they were ad-libbing from a basic script. Several times the filming was cut, for what seemed to be no reason, until I realized it was because the noise from a nearby plane was messing with the sound. The third time this happened, the director got an exasperated look on his face.

Finally the actors gave their most impassioned performances, and it was time to move on to the next scene. Maybe they were just pissed off because they had to do so many takes.

That evening, Kevin, Adam and Alan took us on a whirlwind tour of L.A. nightspots, starting with the hip restaurant-of-the-moment, Le Colonial.

We were seated at a round table near the entrance, so I could watch the people entering. One of them was actress Daphne Zuniga, who played "Jo" on "Melrose Place."

As everyone got lubed up with martinis, we started talking about our personal lives. Adam revealed he had been married many times (he refused to say how many, like it was an anomalous number), and that he had hit rock bottom after doing too many drugs and drinking too much.

Meantime, Kevin continued his unabashed campaign flirting with Mandy, who was the only single woman among us (thank God for small favors).

Kevin was married, but it was evidently an "open relationship." He told us the story of his bizarre LA-style marriage, in which he and his wife lived on opposite ends of their huge mansion, and rarely saw each other. They lived like single people, and Kevin had as many relationships as he wanted.

We had a discussion about LA vs. East Coast marriages. "On the East Coast, you marry for life," proclaimed Kevin. "In L.A., it's until someone better comes along." And evidently sometimes not even that long.

After the meal, Mandy had to make a flight back to the East Coast. "Tee time tomorrow is 9 a.m.," she chirped. I pictured Mandy at some chi-chi golf course, decked out in her immaculate little golf skirt and Tretorns, a white sweater tied perfectly around her shoulders.

Kevin tried to act coy, as if he would not give her the ride he earlier promised to the LAX airport. "Make no mistake," whispered Adam to me, "he is definitely giving her that ride."

After they left, we toured the rest of the restaurant. Nell was determined to find Daphne Zuniga, annoyed that she had missed seeing her at the hostess stand. Every time I saw a thin, dark-haired woman, I would proclaim, "there she is," Nell would get all excited, and it would turn out not to be her. We looked all over, and Daphne must have been let out through a trap door or their celebrity fire escape, because she was nowhere to be found. (What would we have done if we'd found her?).

Then we headed out to "Club Tata" to sample the nightlife. Adam and Alan parked their Mercedes in a parking lot filled with upscale cars. Outside there was a line to get in, but because Kevin's father knew the owner, we slid right in.

While waiting at the bar, a group of nine women strolled by and sat themselves at the center of the room. The Algonquin Silicon Table, if you will. They were all stylishly thin, wore low-cut, revealing gowns, and together added up to more cleavage than a year's worth of Playboy Playmates combined.

When Catherine and I returned from the women's rest room, we commiserated about how our un-nose-jobbed, shoulder-length haired, business-suited appearances put us at downright dowdy on the LA looks scale. While I didn't feel half bad on my native Right Coast, in Club Tata I felt about as sexy as Miss Hathaway from the "Beverly Hillbillies."

Adam glanced over at that table of Breasts that Ate LA, and kept saying, "God bless 'em. God bless 'em." Meantime, we took bets on how far Kevin and Mandy would go on their sojourn to the airport.

***

Kevin finally showed up at a quarter to midnight. By that time the rest of us had gotten restless at the club, and wanted to move on. Actually, everyone but Nell, who wanted to go back to the hotel, which she did. (We later found out she was two months pregnant).

Catherine and I decided to keep moving. How many times do we get to party in LA, we reasoned.

We hadn't realized how much Kevin had been drinking all through dinner and the club, but by the time we got to the after-hours dance bar, he had attained quite a buzz. I tried to slow it down by nursing a Corona, and after a while started losing steam. We danced and drank until the place closed. At some point Adam just disappeared; I guess he got sick of the business meeting/partying combination, even though we hadn't discussed an ounce of business in eight hours.

Kevin didn't stop then, though. He tried to get more beer. "Sorry buddy, bar's closed," said the bartender. Kevin took a hundred-dollar bill out of his wallet and thumped it on the table. "Three more beers," he said again. The bartender paused, picked up the beer, and got us out three Dos Equis.

As we approached the car to ride home, it was clear Alan would be the one driving. He had somehow managed to stay reasonably sober through the whole evening. Thank God for nerds.

I waited to let Catherine get into the back seat, both because I knew Kevin would be slobbering all over whoever was next to him back there, and because I could tell Kevin would probably prefer Catherine's company back there to mine. Poor Catherine had to spend the ride back un-suctioning his tentacles from her torso and deflecting his sloppy kisses.

Kevin leaned forward as we zoomed back to Santa Monica and kept trying to put a new CD in the car's player. He was fumbling all over the front seat and the CD fell to the floor. Finally Alan shrugged, popped the SpaceHog disc into the player, and turned it up as loud as it would go. We rode back to the otherworldly strains of "In the Meantime," a wild tune about aliens interacting with Earthlings. It seemed especially appropriate.

When we were dropped off at the curb in front of the Doubletree Hotel, Kevin got out and gave us long, huggy goodbyes. "You girls...you know what you are?" he slurred.

"No, what?"
"You girls are really...really..."
"Really..." we prodded.
"...really...smart."

***

Nell had her baby eight months later. She received a note from Kevin and his associates: "This just shows it takes longer to do a deal with AOL than it does to create a human life."

Chapter Seven: Cybercity Ghetto

If your life at AOL was rife with meetings, business trips, and partner conferences in the "real world," the other half of what you did took place "out there," in cyberspace. Each producer had his or her own relationship with his or her information providers, forum leaders, or just the odd online denizen whose flame from Hell landed in said producer's in-box in the morning for Terms of Service violations or other red-flag issues.

As a producer who had been with the company for so darn long in comparison to most of the others by the mid-1990s, I watched the drama unfold "out there" as well as behind the corporate walls.

Although most of what I saw was positive stuff – people meeting the man or woman of their dreams, being reunited with long lost love ones, or putting together grassroots organizations for worthy causes, there was undoubtedly some seedy stuff out there.

Ever since the Internet became a household word, horror stories of online creeps preying on unsuspecting innocents and offline crimes linked to online meetings have abounded. But again, as a worker in the trenches of this fascinating industry, I was aware of many more positive occurrences than negative ones.

Still, there were those eerie events, even if they were rarer, that sent a shiver up one's spine. One of the worst news stories I heard about happened relatively early in the Internet Age, which is still not that long ago – in 1996.

Autoerotic Accident or Murder?

In the Fall of 1996, 35-year-old Sharon Lopatka found the right outlet for her previously unexpressed fascination with bondage and torture: chat rooms and e-mail over the Internet. According to The Washington Post, she entered a chat room on Aug. 22 and asked, "want to talk about torturing to death? I have kind of a fascination with torturing 'til death...Of course, I can't speak about it with my family."

Once her fellow bondage correspondents learned she was talking about the real thing, and not just an online fantasy as is the case with many, they backed off. Except for one man: North Carolina Catawba County computer programmer Robert Glass. Glass and Lopatka exchanged "raw, sexual and violent" conversations by email, describing what violent things Glass would do to her.

After about six weeks of such discourse, Lopatka left for North Carolina to meet Glass. She told her husband she was going to visit friends in Georgia, but she left a note for her husband by her computer: "If my body is never retrieved, don't worry. Know that I am at peace."

Lopatka was an entrepreneur who ran several home-based Internet businesses such as a psychic hotline: "Vilado — America's favorite warlock, will cast a spell for you!" But Lopatka's other dealings with the Internet led her down a more sinister path.

One woman, another "bondage and discipline" enthusiast, tried to counsel Lopatka online, but to no avail: "I want to surrender completely. I want to die," were Lopatka's responses.

Sharon Lopatka died of asphyxiation: Glass told investigators it happened while they were having sex, and the rope they were using strangled her.

Paul Greer, the founder of Wave communications, the software Glass used to access the Internet, has some telling remarks on the case; he says he was aware that some of his customers used the sex chat rooms. "You would consider me uninformed to say that's not going on...but so are Mom and Dad communicating with their kids in college, and furniture makers here transferring their data in a spreadsheet to Saudi Arabia, and older folks on the geriatric network...It's a superhighway. It's a new day."

Writes the Washington Post: "That new day brings new challenges for law enforcement and new opportunities for crime. Five years ago, it would have been harder for Robert Glass and Sharon Lopatka to have met. Now, on the exploding Internet, like-minded individuals easily can connect."

***

That has always been what I saw as one of the most powerful and positive features of the Internet: a "global village"-enabler — a vehicle for someone in Des Moines and someone in Rochester, who would otherwise never have found each other, to link up and share their obscure interest in beekeeping or what have you.

Whereas, before, people were limited by geography in whom they dated or befriended, the Internet breaks down those barriers of distance, and in many cases allows people to find the "soul mates" they would never have otherwise.

When I was a little girl, I wondered, as most all little girls do, who my future husband would be. Somewhere out there was my soul mate, and I was bothered by the logical probability that the truest soul mate out there most likely didn't live anywhere near me. Would I have to "settle" for a local boy; for someone I happened to room near in college?

With the Internet and America Online, a much larger world opened up to me. Now there are some inherent dangers in that global village. But you wouldn't stroll Times Square at midnight without taking precautions, and so should you not online.

Steve Case, at the AOL company meeting in November 1996, said AOL was then a village of seven million people...larger than any city in the world. And, he acknowledged, just as you get some bad eggs in real life, a city of that size is naturally going to have some people to watch out for.

Fortunately, on AOL, your tools, such as Parental Controls and Mail Preferences, are much more palatable to the average citizen than carrying a handgun in the downtown district at night.

And too, there are other dangers online services and the Internet enable: that of allowing people to assume other identities; to wear masks.

A Mind-Rape

Offline dangers are just part of what one needs to watch out for in the brave new world of the information superhighway. As the next story illustrates, online chat makes it very easy for one to masquerade as another identity – age, sex, physical appearance, you name it.

One day shortly after I started work at Quantum, my boss, Chris, handed me a photocopy of an article. It was, and still is 15 years later, the strangest story of online relationships I ever read. It was titled "The Strange Case of the Electronic Lover," by Lindsy Van Gelder, and it ran in Ms. Magazine in 1985 (still very early in the days of modems and online chat). It was about a man who used an assumed gender and identity online to engineer his own offline affair, and then pumped the woman with whom he had the affair for information online about himself by using that other identity.

The man, who was a prominent New York psychiatrist (called "Alex" in the story), had created a whole separate online persona for himself: a disabled woman named Joan. "Joan" then introduced Alex, himself, to a woman via CompuServe's CB simulator. CB Simulator was then CompuServe's version of AOL's People Connection, where people communicated by typing on their computers and reading the typed text of others who are connected to the same screen. (This, of course, was before AOL bought CompuServe. In those days, Compuserve was a better-known and larger entity than AOL/Quantum).

"The unfolding of an on-line relationship is unique, combining the thrill of ultra-futuristic technology with the veneration of the written word that informed 19th-century friendships and romances," wrote Van Gelder. "Most people who haven't used the medium have trouble imagining what it's like to connect with other people whose words are wafting across your computer screen. For starters, it's dizzyingly egalitarian, since the most important thing about oneself isn't age, appearance, career success, health, race, gender, sexual preference, accent, or any of the other categories by which we normally judge each other, but one's mind."

Van Gelder goes on to write that she personally often responded to the minds of people who she otherwise might not meet, due to her prejudices or theirs. So the anonymity factor helped in that respect. Handicapped people are as well benefactors of this feature of online communication. They can choose to reveal their disability or not, but they won't get the funny looks, unwanted pity, or avoidance that might saddle their ability to forge relationships in the "real world."

Something about the online means of communicating with someone also expedites the intimacy of a relationship, be it friendship or romance. "For me," writes Van Gelder, "the only odd thing about these relationships has been their chronology. It's a little surreal to know intimate details about someone's childhood before you've ever been out to dinner together."

The "Electronic Lover" case to me is remarkable because of the ingenious machinations Alex used to spin his deception. Because Joan was supposedly handicapped and embarrassingly disfigured from a supposed car accident with a drunk driver, she said she didn't want to see anyone in person. She rarely communicated via telephone, but when she did, she'd "make horrible noises into the receiver – little yelps and moans."

Evidently Alex's alter ego was created by mistake. He was chatting online, using the handle "Shrink, Inc." with a woman who mistook him for a female psychiatrist. He thought the woman was "open with him in a way that stunned him," according to another woman online who was Alex's friend. "What he really found as Joan was that most women opened up to him in a way he had never seen before in all his years of practice." According to this friend, Alex thought he could help them.

"He later told me that his female patients had trouble relating to him – they always seemed to be leaving something out," said another woman who knew both Joan and Alex.

Joan had devised an "almost novelistically detailed biography," including driving clear across Iceland to get over her agoraphobia, having an abortion at age 16, and volunteering with the Manhattan police who patroled for drunk drivers.

Ironically, Joan tried to keep the CB free of impostors. As Van Gelder notes (and as is plenty apparent on AOL if you hang out in People Connection long enough), online chat is rife with impostors, so beware. "There are numerous other gender benders," writes Van Gelder, "some of them gay or bisexual men who come on in female guise to straight men."

Joan was "sexually aggressive" and would pressure her female online friends to engage in "compusex," where, as Van Gelder describes it, "people type out their hottest fantasies while they masturbate."

One of Joan's targets was Janis Goodall, who although being a very close friend of Joan's, told Joan she was heterosexual and not interested. But then Joan suggested Janis meet this "great guy" who was also online, whose name was Alex. After a positive meeting online, Alex sent Janis a round-trip plane ticket to Manhattan and wined and dined her like a queen. They became lovers, although "their sex life was less than satisfactory," supposedly due to Alex's back condition.

When Janis returned home to Berkeley, there was Joan, online, wanting to know if Janis thought she was in love with Alex, and was the sex good?

Several months later, all of "Joan's" online acquaintances started putting two and two together and realized Joan was an impostor. The disabled women felt Joan's extraordinary stories just didn't add up. Eventually Alex was confronted by some of his online friends, and the story got out to the whole circle who knew him.

When Janis was informed in person over the phone by Alex that he and Joan were he same person, she said she went into shock. "I mean, I really freaked out. I wanted to jump off a bridge."

The community of people online who fell under Joan's deceptive spell felt understandably betrayed. But it's not fair to blame the medium. There is a quote in the article from Richard Baker, Compuserve spokesman at the time, who said "Blaming CompuServe for impostors makes about as much sense as blaming the phone company for obscene calls...Our experience has been that electronic impersonators are found out about as quickly as are face-to-face impersonators. While face-to-face impersonators are found out due to appearance, on-line impersonators are found out due to the use of phrases, the way they turn words, and the uncharacteristic thought processes that go into conversing electronically. I also believe that people are angrier when they've been betrayed by an electronic impersonator," he said.

The Phone Call to "Holly" Should Have Tipped Her Off

But my favorite gender-bender 'Net tale is the story of the woman who married another woman, thinking her to be a man.

Margaret Hunter's new husband, Thorne Wesley Jameson Groves, claimed to have AIDS, but never mentioned doctors or medical bills. "His" breasts were covered with heavy bandages, which he said were necessary because of rib injuries from a car wreck.

Hunter met "him" on the Internet. But "he" turned out to be a "she."

Here an excerpt from the clipping, lest you think I am making this up.

ALEXANDRIA, Va. (AP) — It was one thing that Margaret Anne Hunter never saw her new husband undressed, or that no one from his family came to their wedding.

But she stopped believing her husband, Thorne Wesley Jameson Groves, when his parents finally called and asked to speak to "Holly."

Four months after she got married following a relationship that began on the Internet, Ms. Hunter is suing: It turns out her husband was really a woman.

Holly Anne Groves, 26, of Bryan, Texas, pretended to be a man dying of AIDS to avoid sex, Ms. Hunter says in a $575,000 fraud lawsuit.

Throughout their courtship and marriage, Ms. Groves' breasts were covered with heavy bandages, which she said were needed because of rib injuries from a car wreck, the suit said.

Ms. Hunter, 24, of Alexandria, wants reimbursement for money she spent on food, transportation and telephone calls during the relationship.

Although same-sex marriage is illegal in Virginia, Ms. Hunter also wants an annulment "to rest any questions as to her marital status," according to court papers.

She said she no longer uses online services. Computer users "need to be increasingly careful with whom they speak," she said.

***

So how could Ms. Hunter have been fooled? Apparently "Holly had such credible and detailed explanations, excuses and personal history... There was nothing that gave my client or other people pause," said Seth Guggenheim, Ms. Hunter's lawyer.

She did have some clues, however. Although Ms. Groves said she had AIDS, she never mentioned doctors or medical bills.

"He" also claimed to have cancer; an untreatable soft-tissue sarcoma. "He said that his time was short, that his cancer had come out of remission, that he was HIV-positive, that he had only six weeks to live and that he wanted to spend the rest of his time with me," Hunter said. "The feelings that I had for him I believed to be very, very strong."

Ms. Hunter met Ms. Groves electronically in fall 1995. After the two had many online conversations and established a "strong connection," they met in person in December 1995. Groves paid for Hunter to fly to Mexico.

After the two shared a hotel room for two nights, Ms. Groves pledged love and proposed a quick marriage. Ms. Hunter agreed "out of compassion and love," the suit said.

"I saw someone who matched the picture that had previously been sent to me," said Groves. "Someone who looked very much like a man to me."

About 60 friends and relatives attended their wedding at a northern Virginia hotel – "none from Ms. Groves' side."

Although Hunter was suspicious when phone calls and bank checks for "Holly" arrived at their Arlington house, she didn't clear up the mystery until she found a passport and birth certificate belonging to her "husband," reported the Alexandria Journal.

Ms. Hunter was awarded a $264,000 settlement in May 1997 by an Alexandria Circuit Court judge who called the case "one of the most reprehensible matters this court has seen in some time," according to the Journal.

"I asked for punitive damages because I wanted to send a message to Holly and people like Holly that if you commit these kinds of actions against someone it is diabolical, hurtful and it's not right, and that there are serious consequences," Hunter said in her testimony.

"Knowing that he lied not only about his gender, which was terrible, but also about having cancer and HIV was very difficult after I made that commitment to him," Hunter said.

Meeting people over the Internet is especially dangerous, according to Joseph L. Tropea, chairman of the department of sociology at George Washington University, because it skirts the traditional methods through which people acquire information about one another.

"In the past, people very often had kin and family members who could serve as intermediaries and sources of knowledge when introductions occur," Tropea said. "With new technology such as the Internet, those sources of information about others do not exist. As moderns, we may be opening ourselves up to deceit."

Charlie Does Surf

It's not just running into shady characters that you have to wary of online. Among what many would consider one of the downsides of the Internet is its enabling fringe groups and criminals to disseminate information cheaply.

In one example, Charles Manson has his own web site at www.atwa.com. (Apparently putting out t-shirts and music isn't enough for him). The site, maintained by Manson follower Sandra Good, cites its purpose as "to begin to lift the shroud of lies and distortions that have for 27 years been used by self-serving individuals, the mass media, and certain California state departments and offices to cover the reality that is Charles Manson."

The site seems to be devoted to dispelling purported myths about Manson, and promulgating his ideas. For example, under the "Lies" section, several "Lies" about Manson are assailed, such as that Charles Manson is a murderer, Charles Manson was obsessed with the Beatles, Charles Manson wanted to be a rock and roll star, there was a "Manson Family," and that Charles Manson is 5' 2" tall.

A lot of effort seems to be put forth to slay that last one.

The site also debunks the "Helter Skelter" motive of the Tate/LaBianca murders:

"Briefly, this motive is as follows: Manson and his "family" were white racists who hoped to provoke a race war by committing atrocity murders against whites which would be blamed on blacks. The ensuing outrage over these murders would cause whites to retaliate, thus beginning the war ("Helter

Skelter"). While this war was raging Manson and "the family" would be waiting it out in a bottomless pit in Death Valley. (The "bottomless pit", as presented by Bugliosi, is just one aspect of the 'helter skelter' motive which, if truly believed by Manson et. al., would have rendered them psychotic and probably incompetent to stand trial.) The blacks would win the war but not know how to run the world, so they would have to hand the power over to the only white people left on earth: Charles Manson and his "family". Literally fantastic."

Under the hyperlink "Thought," there are inimitable Manson sound-bites such as *"Brother is honor. I will believe in it forever. Without honor we are pus sacks filled with shit."*

Jake Baker: A Case for Free Speech

In 1994, a story broke that was to test the limits of free speech on the Internet. A University of Michigan college sophomore, Abraham Jacob Alkhabaz (who went by the name "Jake Baker" since his parents were divorced five years previously), posted "Gone Fishin'" to the newsgroup alt.sex.stories. The story described "the rape, torture, and murder of a teenage girl and her boyfriend by her brother and his friend." Alt.sex.stories is a newsgroup that, predictably enough, contains amateur writings of a sexual nature. Most of them are of the more soft-porn erotic variety, but Baker's compositions were something different. They were extremely violent, hard-core, and many would say, sick and twisted.

The third story Baker posted, "Doe," on January 9, 1995, was a story which describes the rape, torture, and murder of a woman Baker knew from one of his classes. The story was controversial because some people interpreted it as a threat to this woman. Further, he used her real name in the posted story.

Baker had been "stereotyped in the press as intellectual and socially inept."

The story was for a time posted on the Internet. The guy who put the web page together, PJ Swan, characterizes the piece as "extremely violent and sexual in nature. They are without a doubt both 'indecent' and 'patently offensive.'"

His disclaimer goes on to say:

"Depending on your sensibilities, you may be dismayed or physically nauseated by the material they contain. Don't say you weren't warned. I strongly suggest that you Go Back Now, even if you are not easily upset. They are worse than any "R" rated movie I have ever seen, including Seagal and van Damme movies, and probably worse than most books you have read, including The Silence of the Lambs and books by Anne Rice and Lawrence Sanders. No kidding. Additionally, free-speech advocates may find themselves questioning their views. This is not necessarily a bad thing."

Swan's describes his reaction to the stories:

"I am nauseated by these stories. I would walk out of a movie if it was this graphic. I provide them here as a service to the Internet community regarding the critical issues raised by the Jake Baker case. It is foolish, after all, to argue about something you haven't read. The name of the woman in question has been edited out of these stories and replaced with "Jane Doe"; I do not possess and will not provide her name to anyone. E-mail flames regarding the availability of these stories will be cheerfully ignored. These stories are also freely available elsewhere on the Internet. The stories will go away again if the CDA is upheld by the Supreme Court, so if you want to review them again you may want to save them."

I made an attempt at reading "Doe," and got about eight paragraphs into it before being too grossed out to continue. The story is extremely violent, one could also say misogynistic, and disturbing, but the important question of the case was: did Baker deserve to go to jail for posting it?

The following is the chain of events which led to Baker's indictment:

Baker had also sent e-mail messages to a man in Ontario, Arthur Gonda, describing the kidnapping, rape, and murder of a woman. On January 19 1995, a 16 year-old girl in Moscow read "Doe," and told her father, who told a Michigan alumnus, who then notified the University.

On January 20, DPS officers contact Baker, who waived his Miranda rights and admitted to writing and posting the stories. The officers searched Baker's room and account with his permission, finding an unpublished story and the e-mail conversations with Gonda.

Deemed an immediate threat to the woman named in his story, University President Duderstadt suspended Baker. On February 9, 1995, the FBI arrested Baker on basis of his stories and e-mail to Gonda. Bail was denied on the belief that he was too dangerous to release, as determined by a judge. Baker is charged with violating 18 U.S.C. s 875(c).

Baker was indicted by grand jury on February 15, 1995. He pleaded not guilty at his arraignment two days later. On March 15, 1995, the charge based on the story was dropped, but Baker was charged with five counts based on the e-mail with Gonda.

On June 21, Judge Avern Cohn dismissed the charges against Baker, citing lack of evidence that Baker would act out his fantasies. Although the government filed an appeal of the dismissal, on January 29, 1997, the 6th U.S. Circuit Court of Appeals upheld the dismissal of charges against Baker, ruling that the e-mail messages did not constitute a credible threat.

115

***

All these stories notwithstanding, the online community seems to feel overall that the press has blown the collective perception of the Internet's dangers out of proportion by seizing mostly on these nefarious stories, distorting the reality of the many benefits people receive from this new medium.

"Outside of the trade press, I'd have to say that much of the coverage has taken a rather hysterical tone," says Internet old-timer Vint Cerf. "Much of the reasoning boils down to post hoc ergo propter hoc (event B followed event A so A caused B). The most egregious example of this was the infamous Time cover story that made it seem as if the Internet were nothing more than a showplace for pornography. I, along with many others intimately involved in the development of the Internet, were deeply offended by the narrowness of that story. With the media focusing on those aspects of the Web, it isn't too surprising that Congress passed its ill-considered Communications Decency Act – an incredible example of hysteria inspiring bad law."

Cerf acknowledges that prurient content does exist, and can be an issue for parents to deal with: "This isn't to say that content on the 'net isn't a problem. There is material which is really best-suited to adults only. What one seeks is an ability to enable unfettered global information exchange while providing tools to control access at the periphery for the younger netizens whose parents and teachers rightfully want to make Internet a positive experience for their young charges."

Somewhat ironically, one of the most visible public figures in decrying the breakdown of modern civilization is former model and member of the Gary Hart "Monkey Business" scandal, Donna Rice Hughes.

***

AOL Lawyer Involved in Scandal

It didn't help matters much when a former member of AOL's own staff was implicated in an online-to-offline sex scandal. Attorney Andrew Lewis Singer was charged with sexually assaulting an 11-year-old boy in Loudoun County after learning of his whereabouts via a piece of e-mail exchanged with another boy.

According to The Washington Post, authorities said Singer used the screen name "DCBOY83" to trade instant messages on AOL in early June 1997, then drove from AOL headquarters in Dulles to a pond in Ashburn Farm, Va., where he met up with the 11-year old, put his hand in the boy's pants and fondled him.

An AOL spokeswoman wouldn't comment on the details of the case, except to say Singer had worked for the company about a year and "no longer works for the company."

For its part, AOL has always evinced concern about the dangers of the Internet, and has had "Parental Controls" in place for a long time.

I remember reading newspaper stories hyping sex scandals over the years which failed to even mention the kinds of tools available to parents and others looking to minimize the dangers of the medium for the unsuspecting.

In a January 1998 interview with The Washington Post, Steve Case said the company is deeply committed to the issue. "It's a big deal. We take it very seriously. I mean, you can't not take something like these incidents seriously. We recognize that we're basically a city with 10 million residents, and when you have a city with 10 million residents anywhere — New York, Washington, there might be some activities and behavior that's illegal or inappropriate. And to an extent that goes with the territory. But any one of these incidents is a huge problem. And so what we're doing is saying, 'Look, take a step back here. What can we do to make it a safer environment?'

"And that's been a lot of the focus today, in the past year or so, and much of that came to fruition [with an Internet summit at the White House] where we basically announced a series of programs with AOL.

"One is substantially improving what we call parental controls. They allow parents quite a bit of flexibility, depending on the age of their child, to block some things or everything."

That would suit Fairfax-based anti-pornography organization Enough is Enough just fine. They are urging online companies to employ more people to monitor chat rooms, according to the Post.

But Charles Kennedy, a lawyer who teaches classes at Catholic University Law School on the legal "problem" of the Internet, believes the online services walk a fine line because the First Amendment's free speech protections limit how much they could control online chat. He thinks a better approach is more money and people to track the problem people – the pedophiles who use the Internet.

"Becoming involved with shady characters on the Internet is not so much different than becoming involved with shady characters who are down the block or down the street," he said in The Post. "The Internet is not a uniquely dangerous medium. The danger is that children are unsupervised [there] to an unhealthy extent."

Recognizing that the problem of sexual predators online is a new and growing threat, Oprah Winfrey devoted a whole hour of her program to how parents can protect children online.

The points being made on her program include the prevalence of online smut, and how easily kids can come across it.

One of the quickest ways for kids to become targets for email "spam" (unsolicited email) which contains links to r- or x-rated material is simply to enter a chat room, be it on AOL or a web site. This is because when you are in an AOL chat room, some predators then make note of your email address, or screen name, and use that to send you links in email.

Email Tricks to Get You to Look

The latest insidious trend in email spamming is to give the email the "look" of an old friend or business associate, by phrasing the subject in the form of a question, like "Did you get the package I sent?" The unsuspecting recipient then clicks on the mail, to make sure that it's not something important, and is confronted with an email message like "hot sexy girls!! Click here!"

As Internet safety advocate Donna Rice Hughes points out, "You think somebody is writing you." Someone had sent her an email that was titled, "How'd you find out about that?" She opened the email, and the message said "breasts, legs, full, luscious lips."

Oprah, her friend Gayle King, and Hughes all clicked into the site together, and what they saw was enough to turn their stomachs. "I'm thinking about that egg I had earlier and wishing I hadn't had it. Making me feel a little nauseous right now," said Winfrey.

The scope of pornography on the Internet runs the gamut...you can find Internet newsgroups on every type of sexual deviation, from types of bondage to bestiality. "You can actually see a woman having sex with just about any animal on Noah's Ark, and that is no joke – from snakes to chickens to dogs and horses," says Hughes.

But Hughes is not an all-or-nothing Internet censorship advocate; she seems to understand the incredible positive features of the medium as well. "The Internet has tremendous benefits...we show this not because we think the Internet is bad. But if you don't understand how dangerous it can be, then you may not be inspired and empowered to take the measures that you need to take to protect your kids online."

Another way kids can come across pornography is to stumble onto it when trying to do research for a school project, or when just playing around on a search engine. Suppose they go to a web search engine such as infoseek.com or yahoo.com, and type in a general word for a book report, such as "slavery" – there will be all kinds of hits from that search that include sex-related sites.

Another, more immediate danger to your kids online is child predators. These are people, usually pedophiles, who seek sex-related chat with your child and also may try to lure your child into a location where they can rendezvous offline.

Even if your child has been instructed in avoiding strangers, whether offline or online, these people have very subtle and insidious ways of finding out where to find your child. For example, on the Oprah show, Computer Crime specialist Detective Mike Sullivan of the Naperville, Illinois police department shows how one of these pedophiles might work:

Detective Sullivan started chatting with a teen-age boy from California. They talked about where they live. Minutes into the conversation, he asked the boy for his phone number. At first, he hesitated, but Mike assured him it was OK. And there it was, the phone number. And that's all it took.

Then Detective Sullivan came across the profile of a 14-year old girl who loved soccer. Pretending to be a teen-age girl who plays soccer, too, he struck up a conversation. He asked her for her phone number, but she knew that was dangerous. But what she didn't know was that in this innocent conversation about soccer, she gave a predator all he needs to find her: the name of her school, her grade, and her number on the soccer team.

In another example of an online conversation, Detective Sullivan showed how an everyday conversation between two "teens" about seeing the movies on the weekend could be turned into a road map for a pedophile. He asked a hypothetical teen about which movie theater he would be attending, what time was the show, it's cold, make sure you wear a hat, etc. He said that the boy might volunteer that he'd be wearing his school hat, for example, and bingo, the predator knows exactly where to find the boy and what he will be wearing.

Harassed without a Computer?
The Strange Case of Revenge for a "Hello" in Sidewalk Chalk

Deborah and Mike Bailey's nine-year-old daughter was the victim of an Internet crime, and her family didn't even own a computer! Her parents were featured on the Oprah show because they had received many strange calls from men wanting sex with their daughter. Mike Bailey investigated and found that someone had posted her name, location and phone number on the Internet, with insinuations that she was having sex with her father, and wanted to have sex with other men. It was also indicated that pictures were for sale.

The reason? Evidently their neighbor was so offended by their daughter's scrawling "hello" on his driveway in chalk. "It just goes to show you, you don't even need to own a computer to be the victim of an Internet crime," said Winfrey.

***

So how can parents combat their kids being exposed to this kind of thing? Eventually, Congress may pass more stringent Internet content laws. But until that day, parents need to take a very cautious, hands-on approach. If you are concerned about your kids getting into sites they shouldn't, you can use parental controls to limit what they can access.

One mistake you don't want to make: asking them to install the parental controls themselves. Because some parents feel "stupid" when it comes to computers, they sometimes rely on their kids to set things up for them. But this is one area where you don't want to put the responsibility with them, because they can easily give themselves access to areas you don't want them to have.

Winfrey's friend Gayle King, for example, asked her son to set up the blocking himself: "It's so frustrating as a parent because you'd like to think that you're smarter than they are." Hughes suggests calling the software vendor of the product you're installing, or "ask a friend who knows more about computers than you."

(To use parental controls on AOL, just use keyword "parental controls." You can set limits for e-mail, the web, chat, and IMs).

But there will be times when kids may need to get on the web for educational purposes, for example, when researching a paper, as mentioned above. At these times you can sit down with your child during the time they are actually online. You can save the information they need to use later to the hard drive, or print out hard copies for him or her to review offline.

Remember that your child may get access to the Internet via public library or their school library, too. So they should know what to look out for at those times they won't have the filters you have set up at home.

A basic rule of thumb, according to Hughes, is to keep the computer in a public place in the house, like the family or rec room. "And, talk to your kids about their online activities and friends," says Hughes. Hughes cautions parents to be especially cautious of instant messages.

If your kids like to use chat rooms, it's best to restrict them to the monitored rooms, which have a knowledgeable adult positioned who will be on the lookout for suspicious activity. But that is not a panacea. There was a case where a man disguised himself as a 13-year old in a chat room and met a 14-year old girl, and eventually sent her an airline ticket to visit him in Maine. She is now living with him in Maine and her mom can't get her back, since 14 is the age of consent in that state, according to Hughes.

Tracy O'Connell went through a similar horror when her sister, Ceara, disappeared one December day from a shopping mall with a man she had met online. Her family discovered shredded emails in her trash: love letters between Ceara and Brooker Maltais, a 22-year old she met in a chat room.

The publicity the O'Connells generated paid off four months later, when Maltais was recognized. He was sentenced to four years in prison, and Ceara was returned.

Tracy founded Web Wise Kids, a program that warns parents of the dangers online at the same time as it teaches kids how to make "wise choices on the Internet," according to Winfrey.

But Tracy does not believe that total restriction is the answer. "One of the most important things I tell parents is do not run home and pull the plug from the wall. You'll be robbing your child of a good education and you may be sending them down to Susie's house or Johnny's house or to the library where there is definitely a lack of filtering."

***

But enough about the bad stuff that happens online. No discussion of the Internet would be complete without an acknowledgement of the enormous good it has made possible.

Chapter Eight: An Awesome Tool; A New Way of Life

As I mentioned, thoughout my career in the Internet industry, I've always thought the press focused on the kinkier stuff that happened online, to the exclusion of the positive things. It seemed like the powerful communication engendered by the Internet was getting more of a bad rap than good. Sure it could lead to compromising or dangerous situations, if you weren't careful.

But there has been some coverage of the happier events made possible by the Internet. In one example, on Thursday, January 30, 1996, USA Today reported that the Internet led a boy to discover his estranged mom and mystery past.

The boy, Beau Dugas, had been living with his father since September 1983. Vaughn Arceneaux was supposed to return Beau to his mother, Rebecca Comeaux, as according to a court order. But, said the boy's mother, the two "just fell off the face of the earth." Evidently his father gave Beau the impression his mother abandoned him, because that was how it came out to two women with whom Beau chatted on the Internet.

The women were suspicious of how little he knew about his mother's family, and the women contacted Beau's neighbor, who called the National Center for Missing and Exploited Children (which has a home page on the web at http://141.202.246.92).

Over the years at AOL, we talked about putting together a Missing Children's area a la the milk cartons, but it never came to fruition for whatever reason.

There have been a few cyber-backlash pundits – people like Cliff Stoll, author of "The Cuckoo's Egg," and author Ray Bradbury. Bradbury was quoted in USA Today as saying the Internet can't compete with books. "You've got to be able to take a book to bed with you," he said in a speech in the Silicon Valley. "You can't get that on the Internet. You can't hold the Internet." (This was before palmtops and other handhelds became so popular).

"Online Insider" newsletter author Robert Seidman agrees the media hasn't highlighted the positive aspects of the Internet, but doesn't much blame them: "I won't get on the media for focusing too much on the prurient aspects," says Robert Seidman. "The media does this no matter what the landscape is. As a whole, I think the media has overstated the current usefulness of Internet's capabilities and underplayed the potential."

***

Sexual Revolution

Evidence that the Internet is not "just a fad" comes in a report by psychologists that it is "revolutionizing sexuality." They say it is exposing people to a broader range of contact than they would have otherwise, but also, they warn, enticing some to "quick fixes" that "keep healthy intimacy out of reach," as reported in USA Today.

"The Net is a double-edged sword, and a very powerful one," says Stanford University psychologist Al Cooper. One of the dangers of this new form of communications is the potential to isolate oneself from others, visiting sexually explicit sites, for example. Or they warn that married people might stay in a bad relationship because they have just enough of an outlet online to keep them otherwise satisfied.

These psychologists do cite positives to online life: "If you're a gay man in rural Iowa, if you're disabled, if you're an obese woman who wants to meet men," Cooper says, the Net can find you partners.

[Dr. Ruth Westheimer even weighed in on what she perceived was the inability of the computer to replace the "real thing" when she took a tour of Microsoft's campus in the summer of 1997. She'd concluded that its employees are "overworked and undersexed." "Human relationships and laughter and touching cannot be replaced by any computer," she said, according to the software company's employee newsletter, as reported in the Washington Post "Names and Faces" column. Apparently Dr. Ruth suggested the employees consider the many woodsy places on the Microsoft campus for a "romantic interlude."]

Before AOL went to a flat rate pricing plan, there were many stories of customers who spent an enormous amount of time after discovering the exciting new toy of online chat. Many of them were, it was said, so shocked by their first bill (since AOL then charged by the minute) that they canceled the service.

And one of the early online hosts who worked for us in People Connection reportedly spent all his disability money on AOL. His bills were very high because at the time he had no local access "node" near his town, and he had to dial long distance.

These are the lengths people will go to in order to get their online fix. Of course, the goal of former MTV executive Bob Pittman and other AOL honchos, it's been said, is to get people saying "I Need My AOL," just as "I Want My MTV" was the mantra that weaned the first generation of MTV-watchers in the 80s.

There are actually some legal precedents being set in some cases. One Florida woman lost custody of her children because a judge ruled she was so addicted to the Internet that she neglected them. The court records in the case indicated that after Pam Albridge, 40, separated from her husband, she installed a lock on her bedroom door and spent most of her time on a computer, leaving her 5- and 7-year-old kids to fend for themselves.

Kevin Albridge, her husband, said his wife was, "for want of a better word, 'addicted' to the Internet" and that she ignored the needs of the children.

Circuit Court Judge Jerry Lockett concluded Ms. Albridge's obsession with the Internet had clouded her judgment. She had told the judge she spent a weekend with a man she'd met online, and that he'd asked her to marry him the very first time they met. A psychologist who examined Albridge said that "obsessive computer behavior" is increasingly a factor in disrupting family life, according to the news article. "For some people, it can become like alcohol or sex. It allows us to fantasize, eases our sociability and gets us attention."

In a related case, a woman in Cincinnati allegedly let her three young children "wallow in filth" while she surfed the Internet.

Benefits of USENET Bulletin Boards and the Internet Community

But for most people I know, the Internet has been enormously helpful. I don't think they have time to become addicted.

One man I exchanged e-mail with, an engineer for Ford, said "I'd personally be LOST without USENET. Not so much the alt groups (although I DO lurk in several groups there) as the comp.* hierarchy. I've turned down jobs where I couldn't get unrestricted Net-access. Comp.unix.* and comp.lang.* have saved many a project for me.

"I can post a question and get an answer within 24 hours...sometimes a lot better than real vendors. Of course, it's like asking your brother the mechanic to diagnose your car — you've got to trust him to a point. After that point, it's time to go to a real mechanic."

"I think the best uses of human interaction are the quick and free exchange of information, but that's on the practical front," says Seidman. "The Internet serves as an equalizer of sorts against common discriminations based on race, gender and age. You have to the opportunity to assess someone based on what they are saying, not WHO is saying it. I think this can provide for some very worthwhile exchanges and allow people who might not do so otherwise, to come to know each other. The downside is that as they say, on the Internet, nobody knows you're a dog. You can lie or say whatever you want and hide behind anonymity."

Former Apple executive and author Guy Kawasaki says the best thing about the Internet is the combined knowledge. "I run a mailing list, for example, to gather information to write my business books. The sources that I've developed because of the Internet exceeds anything I could have done myself or pay a researcher to do," he says. "The worst thing is the intolerance people exhibit, often unwittingly, to ideas and opinions that don't agree with their own."

"Perhaps the best thing about the Internet is its enabling power," says "father of the Internet" Vint Cerf. "It can re-connect friends and family in ways no other technology has done. It can facilitate the sharing of knowledge across a remarkable range of interests and it can enable group interactions among geographically dispersed participants. My personal experience is that discovering old friends and making new ones through the Internet is almost a daily occurrence."

He adds the Internet is also proving to be one of the most flexible media available. As it evolves, we are seeing the services of earlier media (telephony, radio, television and print publications) becoming a part of the Internet environment.

"I have learned of heartwarming and lifesaving events carried out through the Internet, ranging from discovering treatment for rare diseases to tracking down long-lost relatives and friends," says Cerf. "The ones that are most interesting involve people meeting online and marrying (but we don't have a lot of data as to how well these marriages last — perhaps the statistics are pretty much the same as for people who met in other ways)."

"I think it is bizarre the number of people who have allegedly sent marriage proposals (complete with rings) to people they've never met face-to-face!" says Seidman. He cites online suicide prevention and things like helping someone find (locate) a lost friend or loved one as "pretty powerful examples of good interaction!"

In some cases, Cerf notes, the Internet has saved lives. "The ones that warm the heart are most are tales of life-saving information being found on the net. In one case, I understand a person having a heart-attack or other fatal episode was online at the time and others on the net were able to direct emergency services to the stricken internaut. In a documentary on PBS, an amateur rock music group that delivered its 'sound' over the Internet got a break when a recording company decided to sign them to do an album. Many school children get their introductions to the Internet by way of the Cyberfair sponsored by MCI, Cisco Systems and Network Solutions. The email we get from students and teachers who discover the creative world of the Web this way is ample reward for a lifetime of effort.

"Of course, there are also stories in which Internet has enabled various kinds of fraud and abuse which are reprehensible, but probably inescapable as the Internet community grows to embrace all elements of our global society," said Cerf.

Longtime Internet user and member of the Internet Engineering Task Force Einud Stefferud echoes these sentiments. "All manner of bad stuff from the rest of society will sooner or later show up on the Internet. As the Internet comes to service a larger and larger fraction of the general population, how could you expect it to not also service all the bad and all the good stuff in society?"

Stefferud says one of the best things about the Internet is that it is a real "Power to The People" kind of phenomenon, "as it is an open market economy, and the Internet just adds value to the global open market economy by lowering the barrier to entry into commerce, and enabling local merchants to cheaply reach global markets," He also cites the support of freedom, as the Internet was involved ("a la UUCP") in the collapse of the Soviet Union and the Chinese Democracy Movement.

"And now how it is supporting 'power to the people' as they find how it offers lowered barriers to entry into commerce. So, it is the way the Internet lowers barriers of all kinds for all kinds of people (including AOL users), for all kinds of interactive purposes," he says.

"Among other things along this line, it has helped to bring many families closer together, especially my own, as my older brother and sister, and their children and ours all get on the net so we now have much more interactive family contact than we have ever had before," says Stefferud.

And what of the Internet's detriments? "Among the most annoying aspects are email spamming, and outright scams," says Cerf. Flaming on newsgroups and email distribution lists are similarly distressing, he adds.

"Although I do not think it either feasible or even sensible to attempt to censor Internet content, I would like to see more tools available to limit access to adult content to adults where this is deemed appropriate by parents, teachers and others entrusted with introducing our children to the Internet and guiding their exploration of its content. The PICS standard for marking web pages, for example, is a useful step in that direction," says Cerf.

Cerf feels that the Internet is a connectivity tool. "It connects us in ways we've never been connected before and that's its power, appeal and its risk. We can see cultures that might have escaped notice becoming visible on the World Wide Web - preserving values and views that might have been lost. As the net spreads, we can see multi-lingual sources of information spontaneously arising. Although much of the Internet's content and transactions are carried out in English (which has become an international language of commerce and research), the net's technology is largely language neutral. We still have work to do to improve the fidelity of languages that need special character sets. Tools to translate among the world's languages will also be helpful in this endeavor," he says.

As a trustee and former president of the Internet Society, Cerf is working toward widespread access to the Internet around the world. "I especially enjoy the contacts I've made with hundreds of school children everywhere who are just beginning to learn how to use the Internet," he says.

"On balance, I think the Internet can be a powerful resource for accomplishing much good in the world and I hope that its users will fulfill that hope by rejecting abuse of this technology and embracing its constructive potential."

Enabling the Physically Challenged

Even the Father of the Internet finds help for a disability: "I have a moderate-to-severe hearing loss (about 65 dB in both ears). Internet and its predecessor, ARPANET, have been of almost incalculable value thanks to electronic mail - a medium which one can characterize as a 'great equalizer.' I have made many new friends and renewed many old acquaintances through its increasing accessibility. I rely on it daily to find information of value to me - often better organized, more up-to-date and in more useful format than I can find in other media."

The disabilities forums on America Online and similar websites have connected people with conditions from Autism to Williams Syndrome, to everything in between.

Career Opportunities

And let's not forget the new careers the Internet has spawned, from webmaster and web page designer to coder: "More generally, with the exception of the time I spent developing MCI Mail, Internet has furnished me with a 25 year career in computer networking," says Cerf.

According to Forrester Research in Cambridge, MA, web site growth doubled in 1997 and nearly tripled in 1998. The size of web content (as measured in megabits of storage) was expected to increase threefold through the end of 1998.

"While the jury is still out on whether the Internet is boosting corporate America's bottom line, the dramatic increase in corporate web sites has created many new jobs," writes Bob Weinstein in The Washington Times.

"Corporate America's love affair with the Internet has created one of the hottest job markets of the decade," says Paul Gavejian, a principal of Buck Consultant's Stanford CT office. One survey found companies devote an average of 14 employees to Internet and web site maintenance, and that they're well compensated for their efforts.

For his part, Stefferud has made the Internet his life since 1975, and has operated a successful Internet consulting practice since then. "It has provided enough for us to retire, given that we live inexpensively and have learned to always live well within our means. Some people think I should 'get another life,' sort of suggesting that the Internet is not adequate to provide a fulfilling life, but I disagree. I think that many people should be as lucky as I to find a career that provided such an interesting lot of work that I can claim that I have really enjoyed almost all my Internet working days. Not always stress-free, but always interesting.

"Actually, as I am gradually retiring, I still find it really interesting to take on a new client for free to make something really interesting happen in the 'Net. I am still active in the IETF, and I intend to continue to contribute as long as I am able to do so," he said.

The Acceleration of Information

Another way the Internet has affected all our lives is in the accelerated dissemination of information. Internet "rogue" journalist Matt Drudge has scooped many established media outlets, in ways that some might say are irresponsible, but none can deny they don't have a strong impact. "After all, it was Internet columnist Matt Drudge – leaking a Newsweek story that the newsmagazine held last weekend – that spurred ABC's Jackie Judd, The Washington Post and Los Angeles Times to break the story Wednesday," wrote Peter Johnson in USA Today when the Monica Lewinsky story broke. It's a far cry from the era when President Kennedy was able to call The New York Times and get it to downplay its story about Cuban exiles training for an invasion, CNN's Jeff Greenfield said.

The Drudge Report, a free e-mail column that is sent out listserv-fashion to its subscribers, is anchored in cyberspace at http://www.drudgereport.com. I had been a subscriber to that list for about a year, when I received the column that broke the Monica Lewinsky story. Here is an excerpt:

CONTROVERSY SWIRLS AROUND TAPES OF FORMER WHITE HOUSE INTERN, AS STARR MOVES IN!

World Exclusive
Must Credit the DRUDGE REPORT

Federal investigators are now in possession of intimate taped conversations of a former White House intern, age 23, discussing details of her alleged sexual relationship with President Clinton, the DRUDGE REPORT has learned.

The tapes were made by a federal employee who has been granted immunity.

According to sources in and out of government, Whitewater independent counsel Kenneth Starr became involved in the situation when he received intelligence that senior administration officials may have offered federal jobs to a young woman in an effort to prevent stories from going public — stories involving sexual episodes that allegedly occurred in a room off the Oval Office.

"Starr is not on the bimbo beat," one source close to the situation told the DRUDGE REPORT late Tuesday. "He's looking at a potential for obstruction of justice charges."

***

Later, Drudge posted an even more salacious leak, one involving allegations of Clinton's semen:

XXXXX DRUDGE REPORT URGENT XXXXX 18:42 UTC WED JAN 21 1998 XXXXX

WATERGATE 1998

WORLD EXCLUSIVE
MUST CREDIT THE DRUDGE REPORT
CONTAINS GRAPHIC DESCRIPTIONS

REPORT: LEWINSKY OFFERED U.N. JOB; INVESTIGATORS: DNA TRAIL MAY EXIST

U.N. AMBASSADOR RICHARDSON OFFERED ME A JOB DURING A BREAKFAST MEETING AT THE WATERGATE HOTEL — WORDS WHITE HOUSE INTERN MONICA LEWINSKY, 24, ALLEGEDLY TOLD PENTAGON WORKER LINDA TRIPP LATE IN DECEMBER 1997.

THE OFFER CAME AS LEWINSKY WAS ASKING TO RETURN TO THE WHITE HOUSE, THE DRUDGE REPORT HAS LEARNED, UNHAPPY IN THE PENTAGON JOB SHE HELD — A JOB THAT SHE STARTED IN APRIL 1996 AFTER BEING RELEASED FROM A WHITE HOUSE POSITION.

"THEY WANTED HER OUT OF THE WHITE HOUSE DURING THE ELECTION," A SOURCE CLOSE TO THE INVESTIGATION TELLS THE DRUDGE REPORT...

SEPARATELY, THE DRUDGE REPORT HAS LEARNED, INVESTIGATORS HAVE BECOME CONVINCED THAT THERE MAY BE A DNA TRAIL THAT COULD CONFIRM PRESIDENT CLINTON'S SEXUAL INVOLVEMENT WITH LEWINSKY, A RELATIONSHIP THAT WAS CAPTURED IN LEWINSKY'S OWN VOICE ON AUDIO TAPE.

TRIPP HAS SHARED WITH INVESTIGATORS A CONVERSATION WHERE LEWINSKY ALLEGEDLY CONFIDED THAT SHE KEPT A GARMENT WITH CLINTON'S DRIED SEMEN ON IT – A GARMENT SHE ALLEGEDLY SAID SHE WOULD NEVER WASH...

DEVELOPING...

****

Currently, Drudge is being sued for libel by White House aide Sidney Blumenthal over statements Drudge made about him in an Internet column. Drudge frequently boasts about how many hits his web site receives from White House personnel, as well as other government workers.

Clearly, the Internet's speed of delivery has eclipsed other media formats in terms of getting the hot news out first.

***

Meantime, the Internet, and particularly the hip, well-marketed America Online, was appealing to the teen generation, sometimes dubbed "Generation Y." One hapless father, Washington Post staff writer Howard Kurtz, complained of his daughter's tying up of the family phone line with her computer chat: "Your household phone becomes as inaccessible as an interview with Monica Lewinsky," he wrote at the time.

"Clearly, we have arrived at an important cultural moment: America Online is the new corner store or shopping mall. It's where the kids hang out. They can gossip with half a dozen friends at once without getting off their rear ends. They determine membership in their clique by placing only the most desirable names on their AOL "Buddy List" [which pops up to tell you when friends are online]," writes Kurtz.

He says that while he relies on e-mail and digital conveniences as "much as the next troglodyte," he remains puzzled at the "magnetic attraction of what AOL calls 'instant messages,' which flash onto one's screen like teletype bulletins, often uninvited." Isn't it better, he asks, to hear people's voices, to hear them laugh, or – this seems quaint, I know – to actually get together and go somewhere?

Just when it seems Kurtz has missed the point, he touches on what makes this form of communication magic, especially for that awkward time of life known as adolescence: it's appealing precisely because you don't have to face the other person; and that makes the communication easier, more intimate.

When he asks his daughters why they spend so much time online, they roll their eyes and say it's "cool." ("Case closed.") "My sense is that the distancing effect of the computer eliminates some of the adolescent awkwardness. They can take an extra moment to compose a witty reply. They can strike a certain pose without having to look anyone in the eye. They use funny screen names and melt into their own secret world," he says.

And he noticed something else; something that begins to unlock the reasons why people spend hours and hours at the computer, and why some blame this nefarious new syndrome known as "Internet addiction" for everything from the breakup of marriages to offline crimes: It's also a way to make new friends. "Judy gets hundreds of instant messages from strangers who noticed her home page for a Lucille Ball fan club – a postmodern shrine to a black-and-white show that first flickered across TV screens in 1951 and featured female characters who schemed face to face against their hapless husbands," he writes.

Kurtz says his kids wander around aimlessly when "no one's on," and then sign on again 20 minutes later. "The other day Bonnie, 12, was typing to a friend and had to get off because I needed that phone line. She was bereft. A few moments later she had an inspiration. 'I could call her,' she said."

***

The easier interaction that comes with online communication extends to other relationships beyond friendships and family, however. It's also facilitating better exchanges between students and teachers, for one. "The anxiety about computers isolating us or making us asocial – the effect here is generally just the opposite," said the dean of New Century College at George Mason University in The Washington Post. "There's an informality in e-mail that encourages closer contact" between students and faculty members, he believes.

One student was able to get an answer via e-mail to a psychology course question the week before her final exam, with explanations to concepts not covered in the book. She said she believed it was the best way to get in touch with a professor.

Another student e-mailed entire papers for feedback to his professors before turning them in for a grade.

So we see the Internet facilitating friendships, information, scientific research, education, and much more. As Internet usage increases, we will no doubt see even more positive effects of this revolutionary technology.

Chapter Nine: Lifestyles of the Rich and Nerdy

The Internet may have engendered much in the way of family communications, scientific research, reunions, and other positive phenomena, but of all the positive fallout of this new medium, the creation of so many "Internet millionaires" is probably the subject on which people most like to focus.

Not everyone who worked in the Internet business gets rich, but AOL was indeed responsible for creating many millionaires.

What do people do when they win the lottery? How does it change them? Being an old-timer at AOL was like knowing a whole bunch of people who just happened to do just that.

If you wanted to make a lot of money in the seventies and eighties, your best bet was to buy real estate and hold onto it. Real estate appreciated by leaps and bounds from 1970-1989.

The nineties were a different story. Real estate depreciated or remained stagnant, and the stock market took off. In particular, many new millionaires were made via high-tech companies who gave stock options to their early employees. Once the companies "went public," or issued an "initial public offering" (IPO) so that its stock would be publicly traded, the options turned into real profit for those employees.

Such was the case with America Online. When I started at AOL in 1988, they handed me a document which spelled out how many "incentive stock options" (ISO's) they were giving me, and at what price. I understood that if the company ever went public, they'd be worth something, but I didn't think about whether that would ever happen. I didn't give it much thought; I just put the document in my work file "just in case" and forgot about it.

But other employees, particularly those who had more options than I did, were not so nonchalant. They also had waited longer to see any booty, and didn't know if they'd eventually see any at all. Some employees would joke about the habit of the technical guru of the company, Marc Seriff, to rally an individual hard-working developer with a statement like "Janet, next Christmas, *aaaaanything* you want!" There were several years when he said this and Christmas didn't yield anything unusual for the employees he regaled with such hopeful statements.

He was also fond of saying riches were "just around the corner." "It's a long corner," one of his employees would joke.

But patience paid off. About four years later, AOLers' wild dreams came true, and in March 1992 the company's stock debuted on the NASDAQ at 11 ½, quickly jumping up to 17.

That morning I woke up excited to go to work, and no wonder: most of what got done that day was standing around in the halls drinking champagne, talking about the stock and what people were going to do with their "found money."

Later, the "underground" AOL employee newspaper, "The Quirk," published "The Going Public Issue." It featured "Vacation Watch," where the supposed getaway plans of key AOL employees were listed, along with joke quotes.

Jim Kimsey's entry read "What hotel do I want to stay at? Doesn't matter. Just make sure it's expensive and the bar is open late. Geez, it's going to be tough finding a vacation spot I haven't visited in the last seven years. Anyone have any ideas?"

Other quotes "from around the company":

- "I've been waiting so long, I'm not sure I can bear to watch it go up and down."

- "I just want to say that I won't forget you after I'm rich. What was your name again?"

- "How many t's in trust fund?"

Of course, not everyone made big bucks on AOL's stock. Some employees who arrived at AOL late in the game were given fewer options at higher prices. The later you arrived at AOL after 1994 or so, the less lucrative the option packages tended to be.

And if you sold all or most of your holdings early on, you would not have benefited from the stock's rise throughout the nineties.

For someone used to scrimping, finally having some money was a great boon. It was interesting to watch your coworkers go through new cars, houses, and in some cases, spouses. One day, when the stock went up to a new high, one of my clients joked that my boss's new home "just got a new wing."

That same boss joked to me as I glided into the AOL parking lot one morning in my spanking new dark-green Camry, complete with "moonroof," leather seats, and V6 engine, "nice options car, Julia."

But the real trophy mobiles were the Jaguars among us. V.P. of Marketing Jan Brandt tooled around in a silver XJS, and Jim Kimsey was known to have had different Jag models, including a sporty convertible. But not everyone had the most luxurious set of wheels. Interestingly, at the time, Steve Case drove a comparatively modest vehicle, content with his several-year-old Infiniti.

Steve did purchase a considerably luxurious home in the tony Northern Virginia suburb of McLean, Virginia. He paid $1.125 million for his new residence – probably the flashiest thing I'd ever heard of him doing. Of course, even the $1.125 million is well below what he could potentially afford. Although

Steve's annual compensation is only a few hundred thousand, his AOL stock was valued at $60 million at the time of the August 1995 profile of him.

Bob Pittman was soon to join Case as a neighbor in the Great Falls, Va. Area...Pittman was reportedly shopping for new digs in that vicinity – a little something in the neighborhood of $1.5 million, according to the Post's "Storming the Castles" story of Dec. 30, 1996.

But if not everyone was trading up to a McMansion, some found the influx of cash removing them from reality of their less-fortunate friends and neighbors. "Why didn't they just pay cash?" asked one high-ranking female AOL executive when hearing of someone's mortgage deal, according to one story I heard.

***

When you can lose tens of thousands of dollars when the a stock moves just one point down (and conversely, gain tens of thousands of dollars when a stock moves just one point up), it tends to make you pay pretty close attention.

In Doug Coupland's novel "Microserfs," based on the life of a fictional Microsoft employee, he writes of how employees at that company are fixated with checking the stock quote several times a day:

"Most staffers peek at WinQuote a few times a day. I mean, if you have 10,000 shares (and tons of staff members have way more) and the stock goes up a buck, you've just made ten grand! But then, if it goes down two dollars, you've just lost twenty grand. It's a real psychic yo-yo."

AOLers checked the stock so much that some programmer finally set up an internal keyword that took you directly to AOL's current stock quote on the service's quote checker, rather than having to use keyword "stock" and then input "AOL" like most people have to do.

The stock was a double-edged sword for some, though. Known as "the golden handcuffs," the stock options made most employees extremely reluctant to ever leave their jobs, however much they hated them. It was beneficial to AOL in many ways, since they were a great incentive for their talent to stay.

But AOL and Microsoft were hardly unique in the success of their stock prices in the nineties. Software and Internet companies flourished and multiplied in this time period, particularly in the "Silicon Valley" area of California where a lot of the start-up Initial Public Offerings (IPO's) originated.

"The Valley's boom is being fueled by the surging technology industry," Julie Schmit wrote in a March 1997 article for USA Today. "Located just south of San Francisco, the Valley in the past two years has seen a 65% jump in new stock offerings, explosive growth among Internet and networking companies and more than a doubling of venture capital funding."

According to the article, "kitchen and bathroom remodeling jobs are so in vogue that contractors are booked 18 months out. Times are so good for one venture capitalist that he decided it was time to move his pool to another part of the yard. Home prices are skyrocketing, up 21% last year to a median of $1.3 million in the CEO haven of Atherton. And golf memberships are must-haves at any price. Membership at the Palo Alto Hills Golf & Country Club now costs $110,000, up from $65,000 two years ago. Cash only."

For a while, it seemed like any company with "Internet" in the name was bound to be a sure success. One AOL alumnus, Sunil Paul, became a multimillionaire when he sold the company he helped found, "Freeloader," a "push technology" software company that created software enabling people to retrieve data from the World Wide Web and view it offline.

"Consider Trevor Stout, 26, and his partner, Randall Schmitz, 28. Last year, they sold their company, Internet Business Solutions of Mountain View, for $2.1 million. The year-old company had just 16 employees. But it also had annual revenue of $1 million and a lead in World Wide Web site development. After the sale, the college pals went on a spree. They bought convertible BMWs and homes in cities that command prices in excess of $400,000. They also founded their Breakthrough Software, which they expect to be an even bigger success," writes Schmit.

The Internet stock hype seemed to peak during the IPO of Netscape in the summer of 1995. The stock was much ballyhooed, and people from all walks were calling the company looking to get into this putatively hot new offering.

A group of AOL employees, followers of this company already, discussed pooling resources to buy a chunk of the IPO stock.

Which Came First, the Nursery or the Girlfriend?

With so many people making so much money at this time, what did AOL employees do with their windfalls? I certainly can't speak for everyone, but anecdotes would circulate among different social circles at the company.

One young man allegedly built a bachelor pad ski retreat near Colorado. Another moved to Nevada to reap the benefit of that state's low taxes.

One of the funnier anecdotes I heard was of a brilliant young software engineer who had a mansion built near Reston, Virginia. It allegedly had a good-sized nursery...even though this young man wasn't even dating anyone at the time. (I guess you have to plan ahead sometimes, though!)

Some people took the bull by the horns and made their dreams come true. One couple, upon hearing from their accountant that they would never have to work again, decided to take a year off and do everything they really wanted to do, including taking a cooking class in Tuscany, and, for her, spending a week at a fancy spa with her mother and sister.

Another friend of mine bought a vineyard with her husband.

Some people did "good works" with their money. Many gave to charity. Steve Case and Jim Kimsey were famous for their philanthropic endeavors. And one former marketing executive decided to be a part-time schoolteacher, as that was something that made him feel he was making a difference.

One of my friends just retired to relative relaxation. He described one of his typical days thusly: Wake up whenever, watch CNBC's "Squawk Box." Go out for a run, winding up at Starbuck's for a coffee break. Then hang out at home and read one of the many books stacked at bedside.

Although not having to sweat mortgages and car notes was sweet, money could have a downside. One young and single ex-AOL programmer would be irritated when his friends would introduce him to girls in bars as "retired," and joke about his being independently wealthy. He didn't want girls to be interested in him for his money alone, of course. One young lady he dated went straight from a relationship to him with another retired AOL employee, one with even more riches. Hmmm.

***

Not everyone retired, though. Some "old-timers" are still at the company. Whether it's that they like watching their longevity "rank" creep up as more and more people cash out and retire, or that they just enjoy working, they opt to stay on in their current position or try out a new one.

And some folks are able to do things that are meaningful to them, but not of the material variety. One ex-AOLer decided to become a substitute English teacher, and another adopted a child from Asia.

Overall, I don't think people turn into shallow materialists, though. I've found that people pretty much stay the same. Most of my friends have had money to buy themselves a few toys, and maybe upgrade their living quarters a bit, but I don't know many conspicuous consumers. They are the same down-to-earth folks they have always been, with the same love of computers, or writing, or art, as before. They just have a little more wherewithal to try to make their dreams come true, or even just to lead the kind of lifestyle they've always preferred.

With the current Internet environment cooling down, web "dot-com's" going "dot-bomb" as the overhyped "Internet bubble" bursts, and stocks plunging or remaining stagnant, the mood is not one of profligate spending. But I don't worry about too many people going broke anytime soon.

Chapter Ten: Leaving AOL: Unlatching the Golden Handcuffs

So if life at "the big A," as I sometimes called it, was so interesting, not to mention profitable, why did I leave it?

Good question. As with many decisions, it was a combination of things. As the company became larger, and there were more and more divisions, people in the "production" world in which I dwelt had more narrowly defined duties.

Whereas before, I was a generalist, who would create an online area from conception to reality to "sunsetting" (phasing the area out of the service, in some cases), now I was dealing with a subset of production issues such as area design, art, and production schedules. And where I had enjoyed the creativity and freedom of starting and running online events in my early producer days, now there was a lot more minutiae and the administrative work of a manager. Hiring, reviews, meetings, and production schedules ruled my days.

Then too, on the home front, by the time my daughter was two, I was finding it increasingly difficult to balance work and home. Keeping up with the email and work treadmill was something I prided myself on, but I came to dread vacations due to the backlog of work that built up.

Once, when I was pregnant with my daughter, my husband and I set off to take a mini-vacation: a four-day weekend in an inn in West Virginia. By Monday, the third day, work issues clogged my mind, including the impending launch of a magazine area, and I opted to return a day early, only to find 99 pieces of email in my inbox, all demanding response or action.

So, with the lion's share of my stock having vested, I decided it was time to try a lower-key kind of life, and a more creative one.

***

One person who edited this book seemed surprised that after all the discussion about online friendships and romances that I did not meet my husband online.

No, we met the traditional way, offline, through a friend of a friend. Even with all the pros of an online relationship, I believe there is no substitute for meeting someone face-to-face, and living geographically close enough to make a close relationship feasible.

But that didn't mean I'd never had a friendship of the online variety.

I did correspond with one guy for two years or so, and finally we arranged to meet offline. He lived in California, so this was no mean feat for someone in Virginia.

But I hopped a plane in June 1992, and met him at LAX airport. He looked a little older than the photo he had sent, but close enough that I knew he hadn't sent me a picture of someone else.

Maybe it was a higher force trying to tell me something, but that night the LA area experienced the most powerful Earthquake in some 20 years, with an epicenter in Big Bear, California.

Earthquake notwithstanding, we had a great time touring Disneyland and L.A.'s sights. But this was not going to be the person I spent the rest of my life with.

Still, I did know many people over the years – online hosts, AOL members, and friends of friends – who met, dated and married the person of their dreams via the online medium.

***

Looking back, I feel incredibly lucky to have been a part of something so revolutionary, and which has come to touch the lives of so many people. It may have been accidental, but I think that I and the many dedicated, bright people I worked with at AOL took a risk in the early years. Some left before AOL hit the big time; others stuck it out. We all left our stamp on the service in some way, I believe; even if it's on an area that has been through several permutations since we left.

I've had a lot more time to spend with my family. But I do miss my "air family." Ted Leonsis used to joke that we, the employees of AOL, would be able to answer our kids with pride when they asked, "What did you do in the war, Daddy [or Mommy]?"

Now AOL is entering a new chapter, with its merger with Time Warner. The Internet is now something everyone can be a part of, from creating their own personal web sites to starting their own businesses. I hope you enjoy making your mark as much as I did mine.

Appendix: A Cyber-Lingo Glossary: How to Speak AOL

"On the whole, I think it is quite fascinating to observe the evolution of language to account for and describe our experiences in the electronic world. Historians and anthropologists of the far future are going to need special dictionaries to deal with the jargon and myriad acronyms our age has spawned."
— Vint Cerf

I was an aficionado of AOL office slang since I started at the company in 1988. I'll guide you to the favored expressions by AOL executives as I remember them evolving, as well as give you a complete primer on "cyber" lingo, along with some background on them where appropriate. You'll be sounding like a bona-fide cyber nerd in no time.

ASCII – pronounced "askee," stands for American Standard Code for Information Interchange. - Basic character set used in almost all present-day computers.

bandwidth – technically, it refers to the transmission capacity of a network; in practice, it's come to mean a human's ability to get things done ("I was just given three more projects; I have no more bandwidth.")

bells and whistles – software programmer lingo for exciting, jazzy features you add to a product or software to make it more exciting and colorful.

beta testing - Testing a pre-release (potentially unreliable) version of a piece of software by making it available to selected users. (Term derives from early 1960s terminology for product cycle checkpoints, first used at IBM but later standard throughout the industry). ("AOL is beta-testing the latest version of their software").

Bmp – short for bitmap. Microsoft Windows graphic file format.

client-server – according to Cerf, has taken on a rather specific meaning - the basic notion was part of Internet's design (in the sense that all computers on the net were considered actors and could take on either the client or server roles or, in some instances, both).

cyber –a prefix meaning relating to the online world; it was coined by science-fiction writer William Gibson (author of "Neuromancer").

cyberpunk – I don't know the exact meaning of this word, but my take on it is a person who wears black, has several body parts pierced, and is into the online world and/or computing. (They also probably read "Wired" Magazine).

download – transferring files from a host system to your own computer; ("I'm downloading my family's history from the genealogy forum.").

double-pointed – (not Spock's ears). When a service or feature is located in more than one place. A similar term is "cross-pointed," when two areas point to each other (for example, finding a handicapped forum in the hearing-impaired forum, and vice versa).

emoticons – typed symbols that are used in email and chat to show emotions, as in the smiley face. :)

get it – to instinctively understand; usually referred to clients or information providers who "get" how to provide an engaging, useful and popular service to AOL's members.

GM – golden master. Term used for when a piece of software is finally finished testing – it is "frozen" and ready to be manufactured.

GUI – pronounced "gooey" - graphical user interface. ("XEROX PARC invented the GUI, but Steve Jobs used the idea for Apple's Macintosh computers, and Microsoft later used the idea to make Windows.").

hack – see kluge (although the term is derived from the term "hacker," this is not necessarily negative).

Information Superhighway – a term coined by Al Gore, who, by the way, did not actually *invent* the Internet.

Internaut — Internet user; one of Vint Cerf's favorite terms.

Jpg – pronounced "jaypeg"; a popular graphics file format.

kluge – (klooj) to hastily piece together a quick but inelegant solution to a problem; a "hack."

learning curve – the amount of time it takes to master something.

leverage – to get the most out of (one could even say exploit).

netizen – a citizen of the Internet. Another Cerf favorite.

no-brainer – easy. "Learning to use America Online is a no-brainer."

offline – not online; of the "outside world." Can also refer to irrelevant discussions in a meeting: "That point's not relevant to this meeting; take it offline after we're done".

push – refers to Internet content being "pushed" or broadcast to the user's desktop. (" 'Push' bothers me because there really is no 'Push,' only automated pull," says Robert Seidman).

schmooze – to rub shoulders with information providers or coworkers.

SnailMail — paper mail, as in that delivered by the U.S. Post Office. Otherwise known as "USnail."

Spam – to flood someone's email box with useless or trivial information. (Especially with ads).

sticky – a service or website so interesting and exciting that users keep coming back, as if they are "stuck" to it.

synergy – mutual benefit from two areas combining or working together. This is one I heard a lot, and it always made me want to vomit. (That doesn't mean I stopped using it).

window of opportunity – the sometimes small amount of time you have to accomplish something, due to other events unfolding that might preclude it. (Microsoft had a small window of opportunity to catch up to Netscape in the browser market. Then Netscape had a small window of opportunity to keep from getting crushed by them).

word of mouse – phrase used to allude to finding out about things via the World Wide Web or email among friends (another Cerf-preferred term).

The Acronyms from Hell

There are acronyms everywhere today, and in every industry. They have insinuated themselves into our daily lives; we worry about DUIs and IUDs, study for our SATs and GREs, get our B.A.s and Ph.D.s, and, if we are lucky, IRAs. We don't so much own a Ford as we own a GTO, TRX or an XJS. Nissan made the 300ZX, or just plain "Z." Some of my favorite rock groups are XTC, INXS, and REM.

But online services probably take the cake when it comes to acronym proliferation – yes, even more than the IRS or the CIA. Because being online currently involves typing words to communicate, to expedite the typing we abbreviate what we're going to say. To that end, here are the most popular online acronyms I've learned over the years:

AFK - away from keyboard

BAK - bak at keyboard

DH – dear hubby (often used in women-oriented communities such as Moms Online)

F2F – face-to-face (often used when referring to an "offline" meeting between cyber pen-pals who have met online and want to meet in the "real world")

FWIW – for what it's worth

GMTA - great minds think alike (usually typed when two or more people type the same thing)

IKWYM – I know what your mean

IM – instant message

IMO - in my opinion

IMHO - in my humble opinion

IRL – in the real world (as opposed to online)

JMHO – just my humble opinion

K - OK

LOL - laughing out loud

OL - online

OMG – oh my God

OTOH – on the other hand

ROFL - rolling on floor laughing

ROFLMAO - rolling on floor laughing my ass off

TTFN - tata for now

TPTB - the powers that be

SAHM - Stay at home mom (used frequently on Moms Online and other parenting sites)

SN – screen name

WYSIWIG – "What you see is what you get." Pronounced "wizzy wig." A document that looks the same when printed out as it does on the screen of your computer, for example, is WYSIWIG.

YMMV – your mileage may vary

***

Smilies and Emoticons

"Smilies" and "emoticons" are rampant throughout message boards and chat on AOL. Although some sneer at them as unprofessional or corny, they are an easy, lighthearted way to express emotion or tone in text that may otherwise be misconstrued. In one online "help" document, AOL described them this way:

"Members use shorthands in this 'faceless' medium to express feelings and show actions or 'body language.' You'll find shorthands used in chat, on message boards, in electronic mail – everywhere!

"Tilting your head to the left or turning your screen on its side will help you to see most of the shorthands. :) <—- See? Two eyes and a mouth!"

When you first enter People Connection (Keyword: PC), most members give a brief greeting, or wave (::waving::) and grin (:D). You can perform just about any action online with online choreography (e.g. "walking over to ClassiLady, giving her a big hug {{ }}"). Or a hug for her can be simply {{{ClassiLady}}}.

Some individuals say that typing in ALL CAPS is considered SHOUTING, so to spare those folks' "ears", you might wish to lower your caps.

Here is a listing of some of the most popular symbols and abbreviations:

:) = smile
:D = smile/laughing/big grin
:* = kiss
;) = wink
:X = my lips are sealed
:P = sticking out tongue
{} = a hug
:(= frown
:'(= crying
O:) = angel
}:> = devil
:D Big Smile, Grin
:* Kiss
:O Amazed
:I Bored
:> Devilish Grin
B) Wearing My Shades
—-<—-@ A Rose
(——U)[Beer
(——Y) Champagne
:-) Humorous, Joking
:-(Sad, Long Face
:-') Tongue In Cheek
:-() Shout
;-) Say No More, Nudge, Nudge
=:-() Scares Me Too
:-! Foot In Mouth
:-$ Put Your Money Where Your Mouth Is
O:-) Don't Blame Me, I'm Innocent
%-/ Don't Blame Me, I'm Hungover
<:-) Don't Blame me, I'm a Dunce
C:-) Blame Me, I'm An Egghead
[:-|] Sent By A Robot
:-)8 Sent By A Gentleman
8:-) Sent By A Little Girl
d:-) I Like To Play Baseball

:-8 I Just Ate A Pickle
>:-) A Little Devil
%-| Been Working All Night
::-) Wears Glasses
:-{} Has A Mustache
}:-(Bull Headed
:-[Vampire
:-# Wears Braces
:-& Tongue Tied
:-D Big Smile, Laugh
C|:-= Charlie Chaplin
=|:-)= Abe Lincoln
:-X My Lips Are Sealed
:-C Really Bummed Out
C=:-) Chef
*<:-) Santa Claus
:-O Mr. Bill
*:o) Bozo
o-) Cyclops
:-)) Double Chin
:-! Bronx Cheer
:-7 Smokes A Pipe
C:# Football Player
:_(Vincent Van Gogh
:-)B Double Chin
:c) Pig Headed
$-) Lotto Fever

Julia L. Wilkinson

Bibliography

Books

Swisher, Kara. *aol.com.* New York: Random House, 1998.

Articles

Brown, Paul B. "Starting Over" (William F. von Meister, entrepreneur). *Inc.*, May 1987, p. 19(2)

Gross, Michael. "The Couple of the Minute." *New York,* July 30, 1990.

"The Online World of Steve Case." *Business Week*, April 15, 1996, p. 78.

Schrage, Michael. "Von Meister's Not-So-Trivial Pursuit." *The Washington Post*, September 23, 1985.

Struck, Doug and Shen, Fern. "A Cyberspace Fantasy Turned Fatally Real." *The Washington Post*, November 3, 1996.

Van Gelder, Lindsy. "The Strange Case of the Electronic Lover." *Ms.*, Oct. 1985, p. 94.

Julia L. Wilkinson

ABOUT THE AUTHOR

Julia Wilkinson worked for AOL from 1988 – 1997, and was Director of Community for womenCONNECT.com 1997-1999. She currently lives in Alexandria, Virginia with her husband, daughter and son.

She has been quoted on National Public Radio, in The Washington Post, and in USA Today; published many articles in industry publications; and been a columnist three times for various magazines and online services.

Ms. Wilkinson was most recently a contributor for Bisnow.com, a web site about high-tech CEOs and their lifestyles; and recently wrote weekly columns on business, technology, and home office issues for the women's business web site womenCONNECT.com.

At AOL, she worked as production manager, responsible for all design and production for four major AOL channels: Life, Styles & Interests, The Newsstand, Hobby Central, and Religion & Beliefs.

In her career as producer, Julia created online areas in conjunction with major media partners such as The New York Times, Scientific American, Hachette Filipacchi magazines, Business Week, Wired, the Army Times, and Windows Magazine; as well as for the News, Personal Finance and Media departments. She designed the Life, Styles & Interests web site and wrote original content for it; and developed original programming, including seasonal features, games, and interactive entertainment for the People Connection area of

AOL's former services Q-Link and PC-Link, as well as for the America Online service.

Her early experience with the interactive channel of AOL, "People Connection," gave her a unique perspective on the stories of people in cyberspace and how dramatically it could affect their lives.

She has written for various publications, including Multimedia Online, Boardwatch, COMPUTE!, CONNECT, ComputorEdge, and Digital Publishing Strategies.

Julia is married to J. Nicholas Gallagher, an attorney. They have a daughter, Lindsay; and a son, Kyle; and live in Alexandria, Virginia.

End